ITC SS 19

Proceedings of the 19th ITC Specialist Seminar on Network Usage and Traffic

October 8-9, 2008
Berlin, Germany

Editors
Anja Feldmann
Matthias Grossglauser
Gunnar Karlsson
Adam Wolisz

Editors:

Anja Feldmann
Deutsche Telekom Laboratories, Germany
Technische Universität Berlin, Germany
Email: anja@net.t-labs.tu-berlin.de

Matthias Grossglauser
Nokia Research Center, Finland
EPFL, Switzerland
Email: matthias.grossglauser@nokia.com

Gunnar Karlsson
Royal Institute of Technology (KTH), Sweden
Email: gk@ee.kth.se

Adam Wolisz
Technische Universität Berlin, Germany
Email: awo@ieee.org

Bibliografische Information der Deutschen Nationalbibliothek

Die Deutsche Nationalbibliothek verzeichnet diese Publikation in der Deutschen Nationalbibliografie; detaillierte bibliografische Daten sind im Internet über http://dnb.d-nb.de abrufbar.

ISBN 978-3-8325-2056-4

Logos Verlag Berlin GmbH
Comeniushof, Gubener Str. 47,
10243 Berlin
Tel.:+4903042851090
Fax:+4903042851092
INTERNET:http://www.logos-verlag.de

ORGANIZED BY

Technische Universität Berlin
and
Deutsche Telekom Laboratories

CO-SPONSORED BY

International Teletraffic Congress
http://www.i-teletraffic.org/

The German Information Technology Society (ITG)
http://www.vde.com/

Nokia Research Center
http://www.research.nokia.com

Deutsche Telekom Laboratories
http://www.laboratories.telekom.com/

Technische Universität Berlin
http://www.tu-berlin.de/

ITC SS 19 COMMITTEE MEMBERS

GENERAL CONFERENCE CHAIRS

Anja Feldmann (Technische Universität Berlin, Deutsche Telekom Laboratories, Germany)
Adam Wolisz (Technische Universität Berlin, Germany)

TECHNICAL PROGRAM COMMITTEE CHAIRS

Matthias Grossglauser (Nokia Research Center, Finland, EPFL, Switzerland)
Gunnar Karlsson (Royal Institute of Technology (KTH), Sweden)

TECHNICAL PROGRAM COMMITTEE

Sem Borst, Eindhoven University of Technology, Netherlands
Prosper Chemouil, France Télécom, France
Song Chong, KAIST, Korea
Costas Courcoubetis, Athens University of Economics and Business, Greece
Daniel Figueiredo, Universidade Federal do Rio de Janeiro, Brazil
Paolo Giaccone, Politecnico di Torino, Italy
Ingemar Kaj, Uppsala Universitet, Schweden
Paul Kühn, Universität Stuttgart, Germany
Michel Mandjes, University of Amsterdam, Netherlands
Petteri Mannersalo, VTT, Finland
Andrew Odlyzko, University of Minnesota, USA
Michał Pióro, Warsaw University of Technology, Poland, Lund University, Sweden
George Polyzos, Athens University of Economics and Business, Greece
Alexandre Proutiere, Microsoft Cambridge, UK
Rudolf Riedi, Rice University, USA
Patrick Thiran, Ecole Polytechnique Federale de Lausanne (EPFL), Switzerland
Piet Van Mieghem, Technische Universteit Delft, Netherlands
Jorma Virtamo, Helsinki University of Technology, Finland
Milan Vojnovic, Microsoft Cambridge, UK
Jan Walrand, University of California, Berkeley, USA
Walter Willinger, AT&T Labs, Inc. - Research, USA

TABLE OF CONTENTS

Preface vii

I. Keynotes

Locating IP Congested Links with Unicast Probes
Patrick Thiran, EPFL, Switzerland 3

Delay Tolerant Bulk Data Transfers on the Internet or How to Book Some Terabytes on "Red-Eye" Bandwidth
Pablo Rodriguez, Telefonica Research, Spain 5

II. Contributions

Revisiting Web Traffic from a DSL Provider Perspective: the Case of YouTube
Louis Plissoneau, Orange Labs, Sophia Antipolis, France
Taoufik En-Najjary, Guillaume Urvoy-Keller, Eurécom, Sophia Antipolis, France 9

On the Relevance of On-Line Traffic Engineering
Bingjie Fu, Steve Uhlig, Delft University of Technology, Netherlands 25

Characterizing and Modeling Multiparty Voice Communication for Multiplayer Games
Gabor Papp, Chris Gauthier Dickey, University of Denver, USA 41

Application of Forecasting Techniques and Control Charts for Traffic Anomaly Detection
Gerhard Münz, Georg Carle, Technische Universität München, Germany 57

On the Utility of Anonymized Flow Traces for Anomaly Detection
Martin Burkhart, Daniela Brauckhoff, Martin May, ETH Zurich, Switzerland 75

Dimensioning of the IP-based UMTS Radio Access Network with DiffServ QoS Support
Xi Li, Carmelita Goerg, Andreas Timm-Giel, University of Bremen, Germany
Wojciech Bigos, Nokia Siemens Networks Sp. z o.o., Poland
Andreas Klug, Nokia Siemens Networks GmbH & Co. KG, München, Germany 91

Locality Analysis of Today's Internet Web Services
Joachim Charzinski, Nokia Siemens Networks, München, Germany 107

III. Invited Talks

Modeling the Evolution of the AS-Level Internet: Integrating Aspects of Traffic, Geography and Economy
Petter Holme, KTH EE/LCN, Stockholm, Sweden 123

TCP Root Cause Analysis Revisited
Ernst Biersack, Eurécom, Sophia Antipolis, France 125

Location and Fairness in Self-Organized Networks
Sonja Buchegger, Deutsche Telekom Laboratories, Germany 127

A Snapshot of 3G Mobile Traffic
Fabio Ricciato, FTW, Austria and University Salerno, Italy 129

Reconstruction of Traffic and Topology from Active Measurements
Gábor Vattay, Eötvös University Budapest and
Hungarian Academy of Science, Hungary 131

Author Index 133

PREFACE

Communication networks are part of our societies' vital infrastructures, and the services they provide are fundamental to our daily work and private lives. The networks are constantly evolving, which cause high uncertainty about current and future network usage and traffic demands. An important trend for the network evolution is that data and wireless networks are increasingly inter-connected with sensor networks, industrial communication systems, as well as ad hoc and femto-cell wireless networks into one big internet. Another important trend is the proliferation of novel application-layer communication and interaction models, including social networks, blogs and micro-blogs, and user-generated content sharing. Many of these models are proliferating on the wired internet and they are increasingly becoming available also on mobile terminals. The traffic patterns resulting from the new interaction models and their dependence on human behavior and social contexts are not well understood, and little is known about the usage of services and the resulting traffic in heterogeneous interconnected network.

Given the importance of communication systems in our society, and the tremendous technical changes that constantly occur, this ITC Specialist Seminar has been aimed at the characterization of network usage and traffic, which is germane to the design, operation and management of all forms of communication networks. The seminar takes place on October 8 to 9, 2008, in Berlin, Germany.

The program consists of talks for seven papers selected from the open call, five invited talks, two keynote talks and a panel. We are honored to have Prof. Patrick Thiran from EPFL, Switzerland, and Dr. Pablo Rodriguez from Telefonica Research, Spain, as keynote speakers. Patrick's talk has the title *Locating IP congested links with unicast probes*, and Pablo's has the title *Delay Tolerant Bulk Data Transfers on the Internet or how to book some terabytes on "red-eye" bandwidth*; the topic of the panel discussion is along the theme of the seminar, with the title *Network Usage and Traffic: Do we need fluid mechanics or plumbing?*

Running a seminar requires dedication and much work from many people. We thank all our devoted colleagues in the technical program committee and the local volunteers who make this seminar possible. In particular we thank Petra Hutt, Heike Klemz, and Britta Liebscher for the local organization and the editorial support. We also thank the ITC International Advisory Council, and its chairman Prosper Chemouil, for their support. Thanks to the German Information Technology Society (ITG) and our sponsors Deutsche Telekom Laboratories and Nokia.

We hope that you enjoy the presentations at the seminar, its proceedings, and the discussions with colleagues in the wonderful city of Berlin.

Anja Feldmann and Adam Wolisz	General Conference Chairs
Matthias Grossglauser and Gunnar Karlsson	Technical Program Committee Chairs

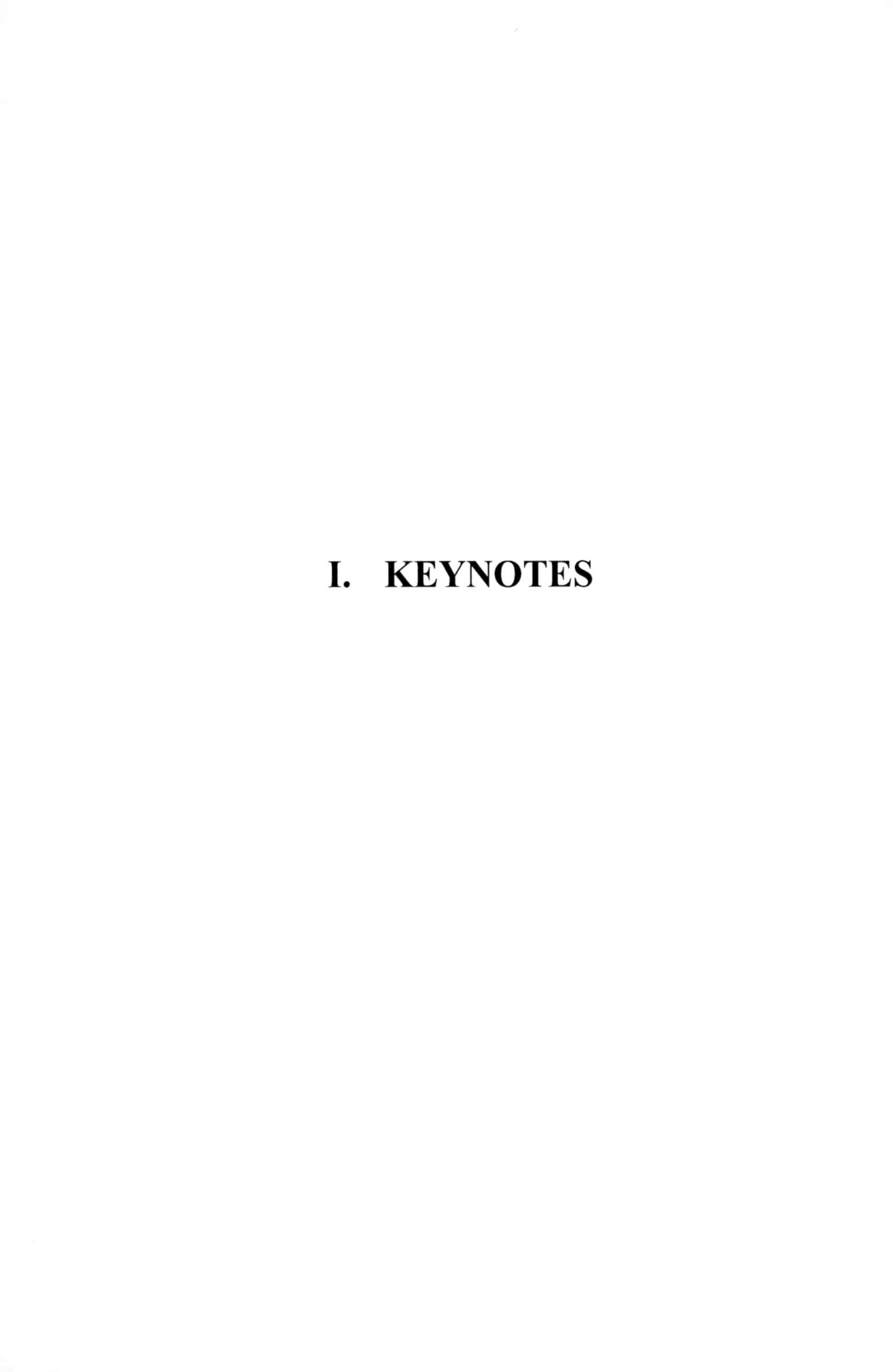

I. KEYNOTES

Patrick THIRAN*

LOCATING IP CONGESTED LINKS WITH UNICAST PROBES

Joint work with Hung X. Nguyen, University of Adelaide.

ABSTRACT

How can we locate congested links in the Internet from end-to-end measurements using active probing mechanisms? As IP multicast is not widely deployed, we want to use only IP unicast probes and avoid relying on tight temporal synchronization between beaconing nodes, which is difficult to achieve between distant sites. Like other problems in network tomography or traffic matrix estimation, this inverse problem is ill-conditioned: the end-of-end measurement outcomes do not allow to uniquely identify the variables representing the status of the IP links. To overcome this critical problem, current methods use the unrealistic assumption that all IP links have the same prior probability of being congested. We find that this assumption is not needed: spatial correlations are sufficient to either learn these probabilities, or to identify the variances of the link loss rates. We can then use the learned probabilities or variances as priors to find rapidly the congested links at any time, with an order of magnitude gain in accuracy over existing algorithms. These solutions scale well and are therefore applicable in today's Internet, as shown by the results obtained both by simulation and real implementation using the PlanetLab network over the Internet.

*EPFL, Switzerland, Patrick.Thiran@epfl.ch

Pablo RODRIGUEZ*

DELAY TOLERANT BULK DATA TRANSFERS ON THE INTERNET OR HOW TO BOOK SOME TERABYTES ON "RED-EYE" BANDWIDTH

ABSTRACT

Many emerging scientific and industrial applications require the capability to transfer large quantities of data, ranging from tens of terabytes to petabytes. Examples include the transport of high definition movies between studios and theaters, and the transport of large quantities of data from telescopes and particle accelerators/colliders to laboratories all around the world. A convenient property of many of these applications is their ability to tolerate delivery delays from a few hours to a few days. Such Delay-Tolerant Bulk (DTB) transfers are currently being serviced through the use of the postal system to transport hard drives and DVDs, or though the use of expensive dedicated networks.

In this talk, based on traffic data from 200+ links of a large transit ISP, we show that the naive approach of using end-to-end (E2E) connection oriented transfers can be prohibitively expensive under widely used 95-percentile charging schemes. We also show that the available bandwidth of E2E connections is subject to time-of-day effects. Based on these observations, we proceed to design a system for performing Store-and-Forward transfer of DTB data and we evaluate the performance of our system under two scenarios: (1) 95-percentile charging, (2) flat-rate charging under time-of-day capacity constraints.

*Telefonica Research, `http://research.tid.es/pablorr/`

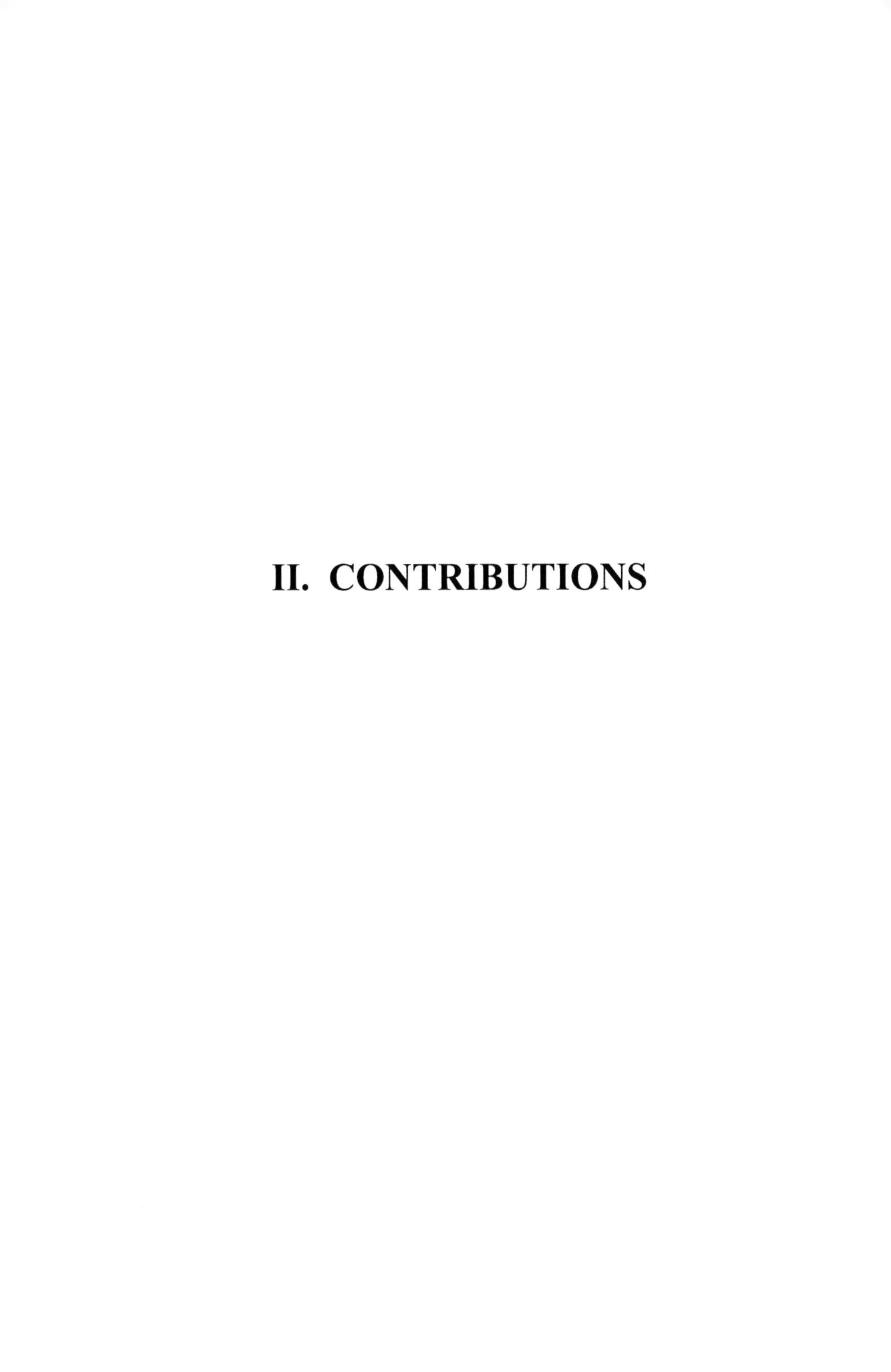

II. CONTRIBUTIONS

Social networks, YouTube, ADSL measurements, Capacity evaluation.

REVISITING WEB TRAFFIC FROM A DSL PROVIDER PERSPECTIVE: THE CASE OF YOUTUBE

Louis PLISSONNEAU*, Taoufik EN-NAJJARY†, Guillaume URVOY-KELLER‡

Video oriented social networks like YouTube have altered the characteristics of Web traffic, as video transfers are carried over the legacy http port 80 using flash technology from Adobe. In this paper, we characterize the impact of YouTube traffic on an ADSL platform of a major ISP in France, connecting about 20,000 users. YouTube is a popular application as about 30% of the users have used this service over the period of observation.

We first observe that YouTube video transfers are faster and larger than other large Web transfers in general. We relate the throughput performance of YouTube Web transfers to the larger capacity of YouTube streaming servers, even though the distribution strategy of YouTube is apparently to cap the throughput of a transfer to a maximum value of approximately 1.25 Mbits/s. We further focus on the cases where the throughputs of YouTube transfers is lower than the playback rate of the video. We relate the bad performance of those transfers to the load on the ADSL platform, thus excluding other root causes like congestion between YouTube streaming servers and the ADSL platform.

Secondly, we focus on YouTube users' behaviors. We have discovered that about 40% of the video transfers were aborted by the client while in 19% of the cases, the client was performing at least one jump action while viewing the video. We show that abortions are only weakly correlated with the throughput achieved during the video transfers, which suggests that the main reason behind a video viewing abortions is the lack of interest for the content, rather than low network throughputs.

1. Introduction

Online Social Networks have become the most popular sites on the Internet, and this allows a large scale study of characteristics of social network graphs [12, 2, 3, 10].

The social networking aspect of the new generation video sharing sites like YouTube and its competitors is the key driving force toward this success, as it provides powerful means of sharing,

*Orange Labs, Sophia-Antipolis, France, Louis.Plissonneau@orange-ftgroup.com
†Eurecom, Sophia-Antipolis, France, Taoufik.En-Najjary@eurecom.fr
‡Eurecom, Sophia-Antipolis, France, Guillaume.Urvoy@eurecom.fr

organizing and finding contents. Understanding the features of YouTube and similar video sharing is crucial to their development and to network engineering.

Recently, YouTube has attracted a lot of attention [5, 11, 4, 9], as it is believed to comprise approximately 20% of all HTTP traffic, and nearly 10% of all traffic in the Internet [1]. Most of these studies rely on crawling for characterizing YouTube video files, popularity and referencing characteristics, and the associated graph connectivity. In [9], the authors have analyzed data traffic of a local campus network. They consider the resources consumed by YouTube traffic as well as the viewing habits of campus users and compare them to traditional Web media streaming workload characteristics.

Our work is along the line of [9] as we focus on actual video transfers from YouTube . Our perspective is however different as we consider residential users connected to an ADSL platform rather than campus users, as in [9]. Our focus is more on the performance perceived by our end users and on determining the root causes of those performance, than on in-depth characterization of YouTube usage.

Our dataset is a 35 hours packet level trace of all traffic on port 80 for our 20,000 ADSL users. Our data collection tool is lossless as compared to the one used in [9]. However, due to privacy constraints, we restricted ourseleves to capture traffic up to the TCP header. We thus do not have access to the meta-data available with YouTube which are exploited in [9].

Our main findings are the following. We first show that YouTube servers are in general much more provisioned than other Web servers servicing large contents. This discrepancy is apparently the reason that explains the significantly better throughputs achieved by YouTube video transfers compared to other large Web transfers in our data. We next focus on the cases where a YouTube transfer is apparently too slow as compared to the video playback rate. We relate the bad performance of those transfers to the load on the ADSL platform, thus excluding other root causes like a bottleneck between YouTube streaming servers and the ADSL platform. Another contribution is to show that transport level information allows to infer the state of a video transfer between a YouTube server and a client. We can thus measure the number of video transfers that are aborted by the client. This allows us to show that it is primarily the lack of interest for the content that motivates abortion of the transfer rather than a low throughput. We also observe that users tend to heavily use the jump facility provided by the Adobe flash player.

2. Dataset

We have collected the traffic of a French regional ADSL Point of Presence (PoP) over a period of 35 hours from 7:20 pm on Thursday 25th October 2007 to 6:00 am on Saturday 27th October 2007. This PoP connects 21,157 users using mainly a DSL box provided by the ISP to connect with contractual access capacities spanning from 512kb/s to 18Mb/s.

In this section, we first present how we detect YouTube video transfers. We next describe our capture tool and the database we use to derive the results in the paper. At last, we present the tool we use to extract client and server side capacity.

2.1. Detecting YouTube Video Transfers

In this section, we will discuss the identification of transfers from the videocaster. Watching a video from YouTube can be done in different ways ranging from browsing the content provider site to following a URL sent by a colleague or watching the video as an embedded object on another Web site, as YouTube offers an API to embed your favorite video in your personal home page or blog.

From a networking perspective, there is not much difference between the above methods to access a given content. In either of the cases, we observe a connection established with a front end server at the videocaster followed by a connection with a streaming server from the company.

We recognized YouTube video traffic based both on reverse DNS look-ups and using the MaxMind database (http://www.maxmind.com/). Indeed, YouTube videos can be provided either by YouTube or Google servers[§], the former being resolvable trough a reverse DNS look-up while the latter are in general not.

A practical issue we had to face was to determine if a large transfer of data from a server in the Google domain (as resolved by Maxmind) is indeed a video transfer for a YouTube video [¶]. In every case (following a URL, clicking on an embedded YouTube object on a non YouTube page or accessing the video through a YouTube Web site), we observed that a connection to a YouTube front end server was done before receiving data from the streaming servers. For every large transfer from a Google machine, we thus checked if the latter was following a connection from the same ADSL IP address to a YouTube server. If it is so, we conclude that this large Google transfer is a video transfer from YouTube.

Using the previous strategy and a threshold of one minute for the look-up of the YouTube connection, we found that all Google transfers of more than 500 kBytes were following YouTube transfers. We chose a threshold of one minute since, as explained in Section 5.1, if the user jumps in the video, a new TCP connection is set up with the streaming server without further interaction with a YouTube server. The fact that all video transfers from Google are initiated after a connection to the YouTube site suggests that the Google video web site is not popular any more, at least on our ADSL platform.

2.2. Capture Description

During 35 hours, users downloaded 1.67 TB of data on port 80. Our capture tool cuts the packets just after the TCP headers and the trace is instantaneously anonymised: the size of our trace in equivalent tcpdump format is 430 GBytes. We show the distribution of volumes (per period of 5 minutes) of HTTP traffic and YouTube traffic downloaded by the clients in Figure 1. The diurnal pattern observed in the trace is characteristic of human activities, as compared to p2p traffic whose volume tends to be more stable over time, e.g. [15, 14]. YouTube traffic accounts for 203 GBytes, i.e., about 12% of the overall port 80 traffic.

From Figure 1, we select 3 charateristic periods for the use of Web:

period A: A high activity period corresponding to the evening of the first day (25th October 19:20pm to 26th October 0:00am);

period B: A moderate activity period from the morning (26th October 9:20) to the end of the afternoon of the second day (26th October 17:20pm);

period C: A high activity period corresponding to the 2nd evening of the trace (from 26th October 17:20pm to 27th October 0:15pm).

[§][9] reported that YouTube transfers could be served by the LimeLight CDN: this scenario was negligible in our dataset.
[¶]Another suspect could be Picasaweb that also allows flash transfers of videos.

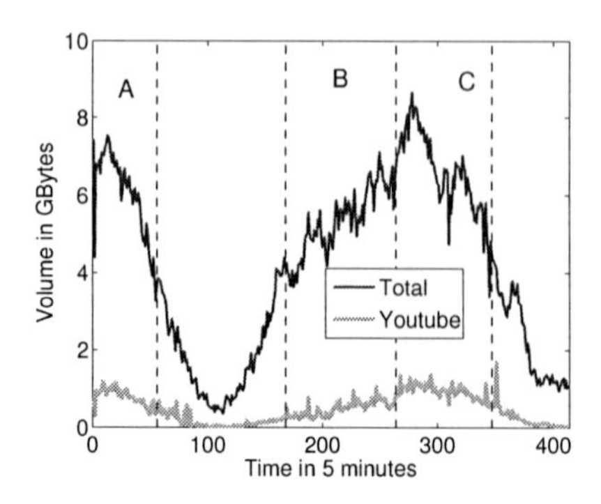

Figure 1: Volume breakdown: Total traffic vs. YouTube traffic over 35 hours

2.3. Database Description

We use a MySQL database to manipulate meta-data concerning each connection: connection identifier, volumes exchanges, throughput, RTT, packet size, reverse-lookup answer and maxmind information, etc... As we declare that a large transfer originating from a Google server is a YouTube video transfer if we observe a connection to a YouTube server prior to this transfer, we end up uploading in the database information about all connections on port 80 of size larger than 500 KB plus all connections from YouTube that are shorter than 500 KB. We have a total of 264,700 long connections in the database, out of which 45,563 are YouTube tranfers. Those YouTube transfers were served by 1,683 servers to 6,085 clients on the platform. The distribution of the number of YouTube transfers per client is given in Figure 2. We distinguish in Figure 2 between transfers in general and transfers that correspond to complete downloads of videos (see Section 5.1 for details). The main message from Figure 2 is that the majority of the clients view only a handful of videos while some are apparently heavy-hitters. We have also introducted in our database the client and server capacity evaluation of each connection (see section 2.4).

2.4. Client and Server Side Capacity Estimation

Since actual capacity may differ from contractual capacity due to attenuation of the line between the customer premise equipments and the DSLAM, we estimate the download capacity of users using a passive capacity estimation tool called PPrate[7]. PPrate is designed to estimate the path capacity from packet inter-arrival times extracted from a TCP connection. We use PPrate as it presents the best compromise among all available passive estimation tools to date (see [8] for detailed comparison).

In PPrate algorithm, the packet inter-arrival times are seen as a time series of packet pair dispersions, which are used to form the bandwidth distribution of the path. As this distribution is multimodal in general, the challenge is to select the mode corresponding to the path capacity. To do so, PPrate estimates first a lower bound of the capacity, and selects as the capacity mode, the strongest and narrowest mode among those larger than the estimated lower bound. The intuition behind this method is that the peak corresponding to the capacity should be one of the dominant peaks of the distribution. Note that the strategy used by PPrate is similar to the one used in Pathrate [6], a popular

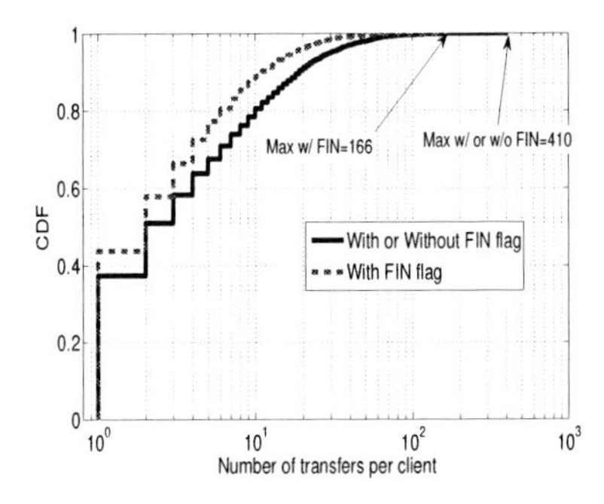

Figure 2: CDF of number of download per client

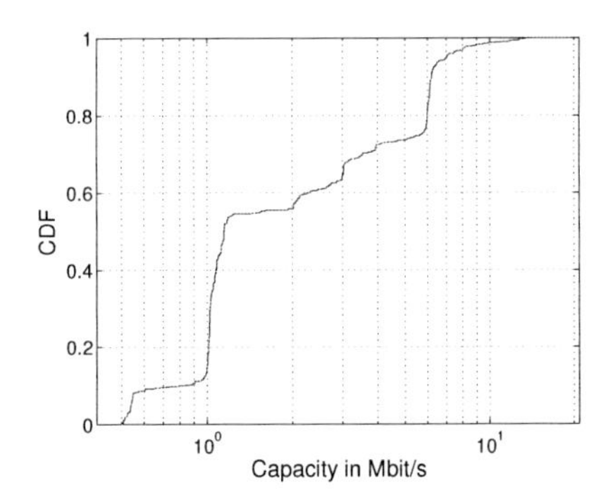

Figure 3: CDF of the Capacity of the ADSL Clients

active capacity estimation tool. More details on the exact algorithms of PPrate, and its comparison with Pathrate, can be found in [7].

Applying PPrate on the TCP data stream received from the HTTP servers, we estimate the capacity of non YouTube and YouTube servers. More precisely, we estimate the capacity of the path between a server and an ADSL client. However, since a lot of Web servers have high speed access to the Internet, and since the core of the Internet is well provisioned, the capacity of the path is in general constrained by the capacity of the server and this is what is measured.

2.4.1. Clients Capacity

In order to get consistent data, we apply PPrate only to ADSL clients having at least 3 YouTube transfers. We consider a capacity estimation for a client as reliable if the various capacity estimates are within 20% of their median value. Figure 2 reveals that about 45% of the clients perform at least 3 YouTube transfers. We eventually obtained a reliable estimation for approximately 30% of the clients.

Figure 3 depicts the capacity estimated by PPrate. We observe peaks at values close to 500 kbits/s, 1 Mbits/s and 6 Mbits/s, which is in concordance with the commercial offers made by the ISP.

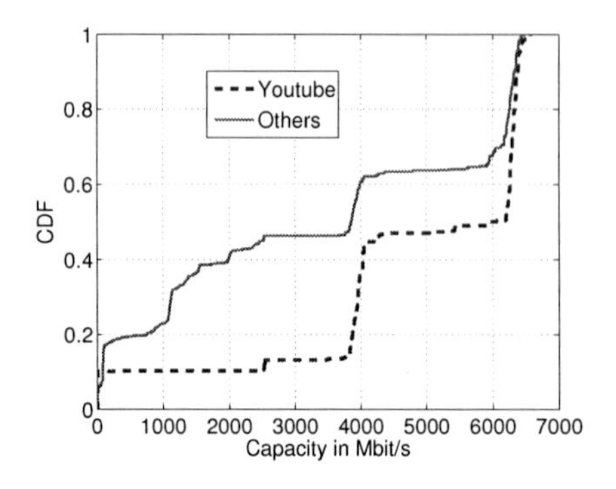

Figure 4: CDF of Servers Capacities: YouTube vs. Others

2.4.2. Servers Capacity

Figure 4 presents the servers' capacities obtained from PPrate. We observe that servers (YouTube and others) are generaly well provisioned (80% of servers have capacities larger than 1Gbit/s). Figure 4 shows that even if Web servers in general are well provisioned, YouTube servers have better access capacity with a significant amount of capacities larger than 4 Gbits/s.

Note that for approximately 5% of servers, PPrate returns abnormally low capacity estimates, i.e. values around 1 Mbit/s. We suspect that PPrate mistakenly picked a peak in the histogram corresponding to the client activity and not to the server activity. Indeed, the estimation technique of PPrate is based on modes in the histogram of inter-arrival times between consecutive TCP packets. Due to the self-clocking nature of TCP, modes exists at values close to the access capacity of the DSL client that the server is currently serving while other modes are closer to the access capacity of the server. It is out of the scope of the paper to detail the PPrate algorithm for choosing the capacity mode (see [7] for details), but in the case of high speed server serving a low speed client, we observed that if an estimation error occurs, it can lead to a severe underestimation of the capacity (overestimation is less likely to occur).

3. Large Web transfers

3.1. Global characteristics

In Figures 5 and 6, we depict the distributions of volumes and throughputs of YouTube transfers against other Web transfers for connections of more than 500 kBytes. The objective is to see how YouTube alters the characteristics of large Web transfers in the Internet.

From Figure 5, we observe that 90% of YouTube transfers are larger than non-YouTube ones. As for the throughput (Figure 6), about 70% of YouTube transfers are faster than non-YouTube ones. This is in concordance with the capacity estimation of the servers (Figure 4).

Overall, we observe that the characteristics of YouTube traffic significantly differ from other Web traffic (for transfers more than 500 kBytes). We could have expected a higher discrepancy since there is a lot of other video transfers in the remaining Web traffic (like Dailymotion or some content

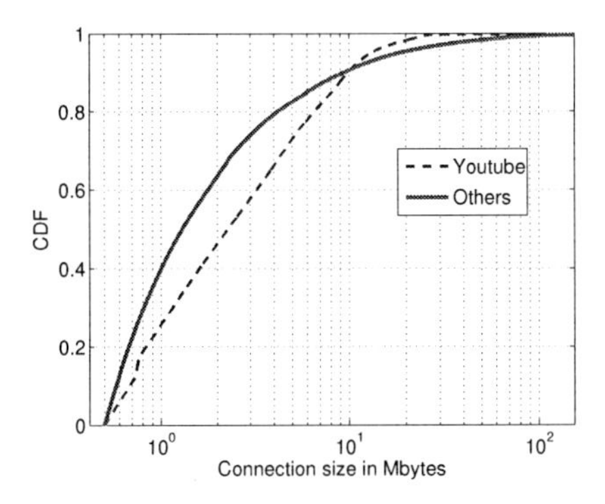

Figure 5: CDF of the Size of YouTube Transfers vs. Others

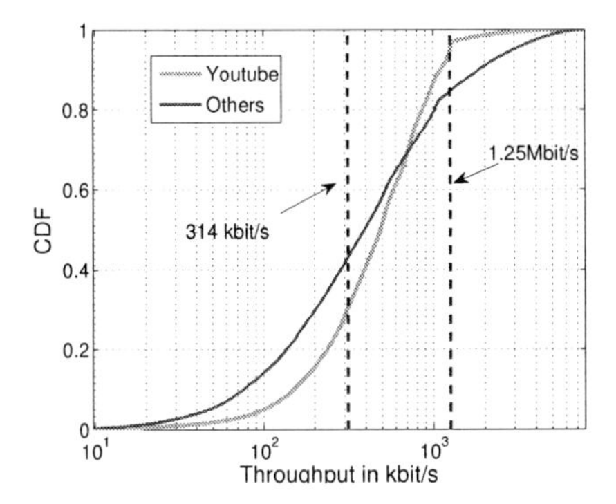

Figure 6: CDF of the Throughput of YouTube Transfers vs. Others

distribution networks like Akamai).

3.2. YouTube Distribution Policy

To understand the distribution strategy of web services, we depict in Figures 7(a) and 7(b) scatterplots of the ratio of the achieved throughput T of a transfer over the clients' capacity C versus the clients' capacity for YouTube and non-YouTube connections respectively. We compute these ratios for connections with more than 500 kBytes of data only and for clients for which PPrate was able to estimate the downlink capacity. A ratio close to 1 indicates that the throughput of the transfer is close to the downlink capacity of the client.

Figures 7(a) and 7(b) show very different characteristics for transfers from YouTube and from non-YouTube servers. For YouTube connections, this ratio is smaller than a specific throughput over capacity ratio in about 96% of the cases. For each transfer x, $\frac{C(x)}{T(x)} \leq \frac{K}{T(x)}$. At the limit: $\frac{C_{max}}{T(x)} = \frac{K}{T(x)} \Rightarrow C_{max} = K$. Figure 7(a) gives an approximation of the maximum throughput for a given

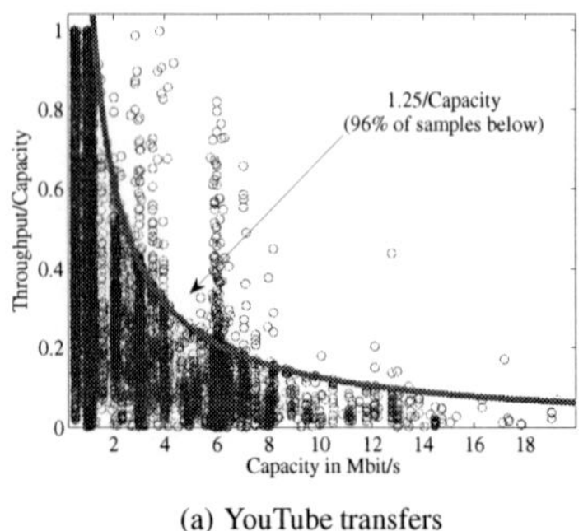

(a) YouTube transfers

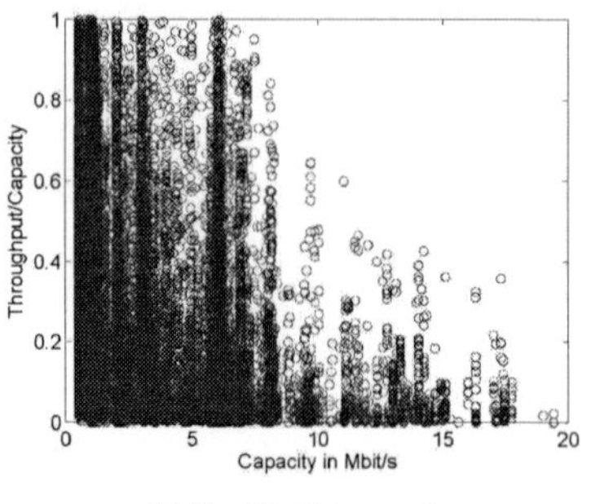

(b) Non-YouTube transfers

Figure 7: Scatterplot of Achieved Throughput over Estimated Capacity vs. Estimated Capacity

YouTube transfer of $C_{max} \approx 1.25\ Mbits/s$. Note that half of our clients have an access capacity smaller than this threshold (see Fig. 3). Even if we restrict to clients with capacity larger than 1.25 Mbits/s, we still obtain that 92% of them are below the above asymptote. Such a distribution strategy makes sense as the playback rate of 97% of the videos is below 1 Mbits/s [9]. In addition, it can prevent ADSL or cable clients with high capacity to consume too much of the capacity of YouTube data centers.

No such throughput limitation is clearly visible for other large Web transfers, as can be seen in Figure 7(b). We simply observe that the higher the capacity of the client, the less likely it is for the transfer to saturate the downlink. One could argue that the latter result is understandable as other large Web transfers come from a variety of Web servers under the control of widely different organizations. However, what Figure 7(b) demonstrates is that the phenomenon observed for YouTube is not an artefact of the ADSL platform we consider in this study.

4. Troubleshooting User's Performance

In this section, we investigate the root cause of the low throughputs of some YouTube transfers. Indeed, we observe from Figure 6 that about 30% of YouTube transfers have a throughput lower than 314 kbit/s. Display of the videos corresponding to those tranfers might lead to periods where the video is frozen, as the vast majority of videos have bitrates between 300 and 400 kbits/s (see [9], Figure 10). Figure 7(a) further suggests that bad performance can occur even to all clients, almost irrespectively of their access capacity.

Several factors can explain why a given TCP transfer achieves a given TCP throughput: the application on top of TCP or factors that affect the loss rate or the RTT of the transfer, as highlighed by the TCP throughput formula [13]. Similarly to the approach followed in [16], we checked on a few example transfers that the application (flash server) was not responsible for the observed low throughput. Indeed, the fraction of push flag (set by the application to indicate that there is no more data to transfer) is negligible for YouTube transfers and packets inter-spacing is never commensurate to the RTT of the connection. It is demonstrated in [16] that those two effects are the two possible footprints left by the application at the packet level.

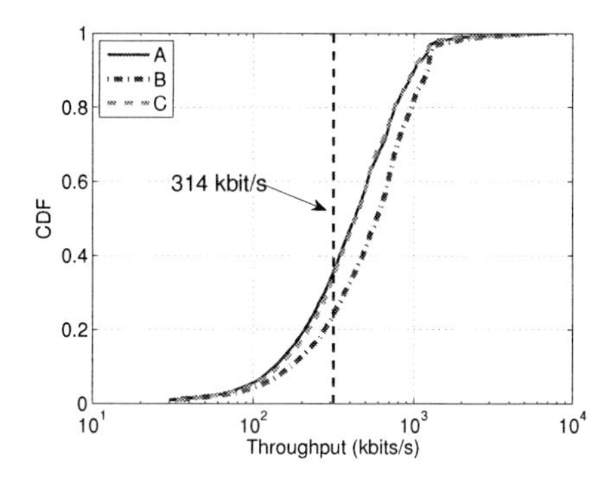

Figure 8: CDF of the achieved throughput for the three periods

We thus conclude that the application is not responsible for the slow transfers. We next focus on the loss rate or the RTT of the transfers. Our intuition is that this is the load on the ADSL platform that explains the low transfer rates we observe. To prove our intuition, we make use of the three periods A,B and C defined in Section 2.2: periods A and C correspond to evenings, and are more loaded than period B that corresponds to the middle of the day. In Figure 8, we observe that the higher throughputs occur in period B, i.e., during the period where the ADSL users generate the lowest amount of Web traffic, which suggests a correlation between the local load and the troughput of the YouTube transfers.

We next focus on losses. The metric we consider, as an estimate of the loss rate, is the fraction of retransmitted packets. In Fig. 9, the retransmission rate is indeed much lower for period B (Note that as we use a logarithmic scale, the mass at zero is not directly visible but corresponds the onset of the curves).

This discrepancy between the periods, in terms of losses, might explain the lower throughputs we observe, but we would like to understand if they are due to a local or distant congestion. We thus consider the RTTs, as increasing loss rates are often correlated with increasing RTT. Our measurement probe, which is located close to the ADSL users enables us to compute the local RTT, between the probe and the ADSL host and the distant RTT between the probe and the YouTube server. Figure 10 (resp. 11) depicts the distant (resp. local) RTT for all YouTube transfers. We observe from Figures 11 that the most likely cause to explain performance degradation of YouTube transfers is an increase of the local RTT, i.e. an increase of the local load. In contrast, distant RTTs - see Figure and 10- seem unaffected by the exact time period one considers.

As a conclusion, we observe that the performance of YouTube transfers is apparently correlated with the local load of the ADSL platform.

5. User behavior

In this section, we present our findings regarding the way users watch videos and its impact on the network traffic. A lot of works, e.g. [4], advocate the use of caching for YouTube and other social

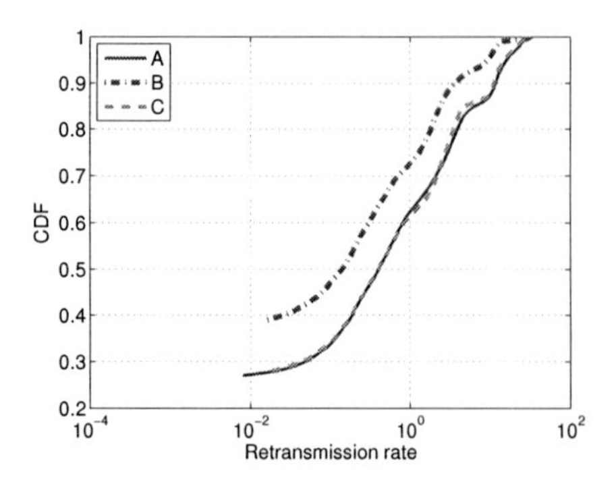

Figure 9: CDF of the retransmission rate for the three periods

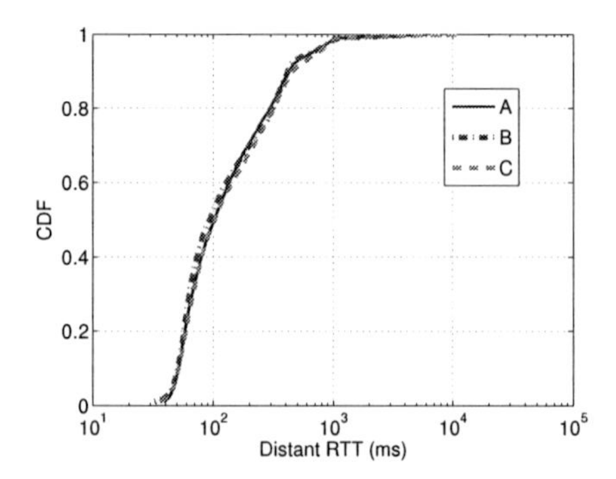

Figure 10: CDF of the distant RTT for the three periods

network traffic. However, while the popularity of videos is displayed by YouTube on its page[||], little is known about the way users watch those videos, and especially if they watch the full video or not. In the extreme case, we could have a popular video that is very long with users watching only the beginning of the video. In this case, caching the full video does not necessarily make sense.

For the above case, we have first performed experimental transfers to assess the state of a video connection and we have then applied our findings to our ADSL trace.

5.1. Assessing the state of a video transfer

Here, we focus on the termination of TCP connections induced by video transfers. Indeed we expect that the end of a TCP connection from the streaming server reflects the status of the file transfer from a user's perspective. We consider the following user behaviors:

[||]See `http://www.masternewmedia.org/news/2008/02/29/internet_video_metrics_when_a.htm` for some details about how videocasters count an actual view of the video.

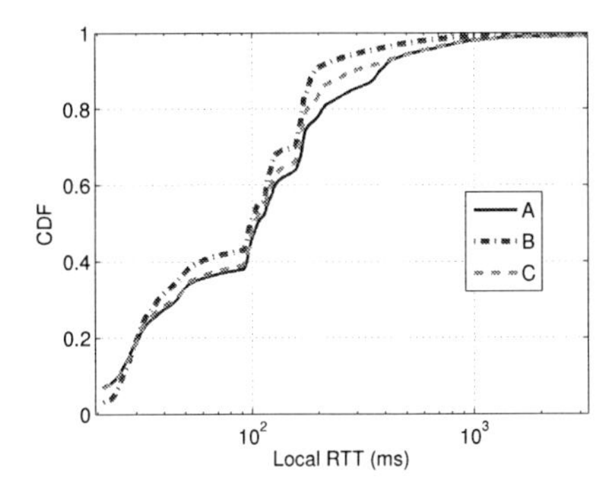

Figure 11: CDF of the Local RTT for the three periods

- Watching a whole video;
- Jumping in a video;
- Switching from one video to another;
- Closing the browser while watching a video.

For the above user actions, we have tested the behavior of the servers using Firefox (Version 2.0.0.12) and Internet Explorer (Version 7.0.5730.13) on Windows XP SP2. The first step of our analysis was to extract the TCP connection corresponding to the actual video transfer from the others. In the case of YouTube, we look for the message `HTTP:GET /get_video?video_id=` ··· In some cases, a message
`HTTP:GET /videodownload?secureurl=` ··· occurs for a video transfer.

At the server side, we have observed only two different types of activity for all user behaviors we have considered. If the server completes the transfer of all data associated with the video, its last data packet will carry the `FIN, PSH` and `ACK` flags. If not, the server will continue sending data as long as it has not received a RST from the client. It then abruptly stops sending data. This also holds true for a jump action that results in the opening of new TCP connection with the same streaming server.

From the client perspective, we observed that whenever the user closes the browser/tab or switches to another video or jumps in the same video, Internet Explorer sends a RST followed by additional RSTs for every new arriving packet from the server. Firefox behaves slightly differently in that it first sends a FIN packet, that is acknowledged by the server but somehow gets ignored as the latter continues sending data that trigger RSTs at the sender side. Eventually whenever RST are received at the server side, the transfer gets aborted.

From the above analysis, we decided to categorize YouTube video transfers into two sets. The first set corresponding to a completed connection with a `FIN` sent by the server and the second set corresponding to partial viewings of videos, either because of the abortion of the viewing or because of a jump action.

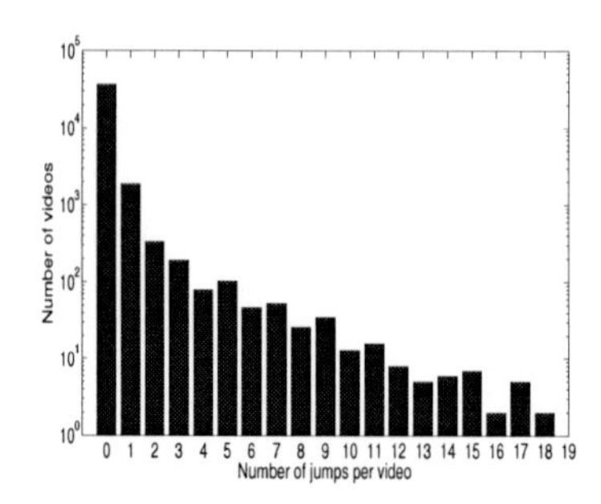

Figure 12: Histogram of jump actions

5.2. When is a view a view?

Using the results obtained from the previous section, we estimated the number of video transfers for which there was a single data transfer between a YouTube server and a client. Over a total of about 45,000 interactions between a streaming server and a client, 36,700 correspond to a single connection while the rest correspond to multiple connections between the client and the same streaming server, i.e. jumps while viewing the video. Note that the actual number of jump actions might be higher than the one we observe as we can detect a jump only if it has an impact on the transfer on the wire. However, in a number of cases, and especially if the video is short, it is likely that the video will be fully received before the jump is performed. In this case, this action is handled by the flash player and has no effect on the network traffic that we could measure.

Let us first concentrate on the 36,700 cases where there is a single transfer. In half of the cases only, the connection terminates correctly while in the other cases, the user stops the transfer either because of a lack of interest in the content or because of bad network conditions. We address this question in Section 5.3.

As for the jump action, we present in Figure 12 the distribution of the number of jumps per video. In 93% of the cases, there is single jump action (note the logarithmic scale of the figure). Not suprisingly, the larger the number of jumps, the smaller the number of samples.

# of jump actions	Mean Volume in Mbytes
0	4.74
1	4.77
2	6.06
3	6.28
4	6.37
5	6.97
6	8.16
7	7.77
8	8.94
9 or more	15.20

Table 1: Mean Volume of Transfers vs. Number of jump actions

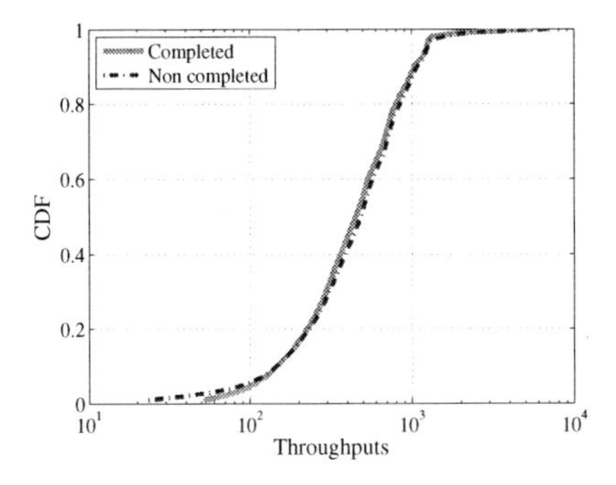

Figure 13: Througputs for completed and non completed video transfers

We have computed - see Table 1 - the total amount of data exchanged between a given client and a given YouTube server, depending on the number of jump actions taken. Table 1 shows that the video transfers with jump actions are much longer than others. Indeed, to be able to jump in a video before completing its download, it must be long enough.

5.3. Video transfer status vs. throughput?

In this section, we focus on the correlation between the status of a video transfer (completed or not), and its throughput. Indeed, there might be two possible causes for the abortion of a transfer: a lack of interest for the content or a too low transfer rate. To investigate this issue, we plot in Figure 13 the throughput of completed video transfers and non completed video transfers. We observe that even if a few percents of the non completed transfers have lower throughputs than completed ones, throughputs in both categories are similar. This suggests that a low throughput is not the primary reason for aborting a video transfer. This is further confirmed by the fact that about 32% of the completed video transfers were performed at rates lower than 314 kbits/s. This confirms that users are quite patient if they really want to view a content.

6. Conclusion

In this paper, we have investigated the characteristics of YouTube traffic. Our main findings can be summarized as follows. YouTube traffic accounts for as much as 20% of the Web traffic. The characteristics of YouTube traffic (volumes, throughputs) significantly differ from other Web traffic when considering large connections (more than 500 kBytes). YouTube servers apply a rate limitation policy for content distribution, with a maximum transfer rate of about 1.25 Mbits/s. This holds even though their servers seem to have better access capacity than an average Web Server in the Internet. As for the impact of user behavior, we have found that about half of the video transfers were aborted, probably because of lack of interest in the content or because of poor network conditions. We have also detected that users perform some jump actions in 19% of the cases.

As an extension of this work, we plan to figure out the observed characteristics of YouTube on a long term analysis, and to compare YouTube traffic to other video streaming ones. We have already established that in the analyzed data, Dailymotion and its content provider LimeLight represent a significant traffic volume in our dataset.

[1] "Ellacoya data shows web traffic overtakes Peer-to-Peers as largest percentage of bandwidth on the network", http://www.ellacoya.com/news/pdf/2007/ NXTcommEllacoyaMediaAlert.pdf.

[2] L. A. Adamic, O. Buyukkokten, , and E. Adar, "A social network caught in the Web.", In *First Monday*, 2003.

[3] L. Backstrom, D. Huttenlocher, J. Kleinberg, and X. Lan, "Group formation in large social networks: membership, growth, and evolution", In *KDD '06: Proceedings of the 12th ACM SIGKDD international conference on Knowledge discovery and data mining*, New York, NY, USA, 2006.

[4] M. Cha, H. Kwak, P. Rodriguez, Y.-Y. Ahn, and S. Moon, "I tube, you tube, everybody tubes: analyzing the world's largest user generated content video system", In *IMC '07: Proceedings of the 7th ACM SIGCOMM conference on Internet measurement*, New York, NY, USA, 2007, ACM.

[5] X. Cheng, C. Dale, , and J. Liu, "Understanding the Characteristics of Internet Short Video Sharing: YouTube as a Case Study.", arXiv:0707.3670v1, Cornell University, July 2007.

[6] C. Dovrolis, P. Ramanathan, and D. Moore, "What Do Packet Dispersion Techniques Measure?", In *Proc. of IEEE INFOCOM'01*, pp. 905–914, Los Alamitos, CA, April 2001.

[7] T. En-Najjary and G. Urvoy-Keller, "PPrate: A Passive Capacity Estimation", In *IEEE e2emon*, Vanvouver, CANADA, 2006.

[8] T. En-Najjary and G. Urvoy-Keller, "Passive Capacity Estimation: Comparison of Existing Tools", In *International Symposium on Performance Evaluation of Computer and Telecommunication Systems*, Edinburgh, UK, 2008.

[9] P. Gill, M. Arlitt, Z. Li, and A. Mahanti, "Youtube traffic characterization: a view from the edge", In *IMC '07: Proceedings of the 7th ACM SIGCOMM conference on Internet measurement*, pp. 15–28, New York, NY, USA, 2007, ACM.

[10] M. Girvan and M. E. J. Newman, "Community structure in social and biological networks.", In *Proceedings of the National Academy of Sciences (PNAS)*, pp. 7821–7826, 2002.

[11] M. J. Halvey and M. T. Keane, "Exploring social dynamics in online media sharing", In *WWW '07: Proceedings of the 16th international conference on World Wide Web*, New York, NY, USA, 2007, ACM.

[12] A. Mislove, M. Marcon, K. P. Gummadi, P. Druschel, and B. Bhattacharjee, "Measurement and analysis of online social networks", In *IMC '07: Proceedings of the 7th ACM SIGCOMM conference on Internet measurement*, pp. 29–42, New York, NY, USA, 2007.

[13] J. Padhye, V. Firoiu, D. Towsley, and J. Kurose, "Modeling TCP Throughput: A Simple Model and its Empirical Validation", In *Proc. of ACM SIGCOMM'98*, pp. 303–314, Vancouver, Canada, August 1998.

[14] L. Plissonneau, J.-L. Costeux, and P. Brown, "Analysis of Peer-to-Peer Traffic on ADSL", In *6th International Workshop on Passive and Active Network Measurement*, pp. 69–82, January 2005.

[15] M. Siekkinen, D. Collange, G. Urvoy-Keller, and E. W. Biersack, "Performance Limitations of ADSL Users: A Case Study", In *Proc. Passive and Active Measurement: PAM 2007*, April 2007.

[16] M. Siekkinen, G. Urvoy-Keller, and E. W. Biersack, "On the Interaction Between Internet Applications and TCP", In *Proc. 20th International Teletraffic Congress: ITC*, June 2007.

Key words: traffic engineering, network provisioning, network design

BINGJIE FU, STEVE UHLIG*

ON THE RELEVANCE OF ON-LINE TRAFFIC ENGINEERING

The evaluation of dynamic Traffic Engineering (TE) algorithms is usually carried out using some specific network(s), traffic pattern(s) and traffic engineering objective(s). As the behavior of a TE algorithm is a consequence of the interactions between the network, the traffic demand and the algorithm itself, the relevance of TE may depend on several network design aspects.

In this paper, we evaluate well-known TE algorithms using real-world and generated network topologies and traffic demands. By re-scaling observed traffic demands, we are able to observe the behavior of TE algorithms under a variety of situations, which may not be observable in reality. We identify distinct network load regimes that correspond to different behaviors of the TE algorithms. We also study the impact of several network design aspects, like network provisioning and redundancy, on the relevance of TE algorithms. We find that there are specific situations under which TE algorithms are useful. These situations depend highly on shortcomings in the network provisioning as well as on the availability of alternative paths in the network.

1. INTRODUCTION

The evolution of the Internet has been amazing in the last decades. New applications with different service constraints have emerged, such as real-time video conference, on-line gaming, etc. High-speed networks are expected to support a wide variety of these sensitive applications. However, the current Internet architecture offers mainly a best-effort service, so providing Quality of Service (QoS) guarantees in the current Internet is not easy. More demanding applications as well as stricter SLA's are driving Internet Service Providers to rely on traffic engineering [1] to better control the flow of IP packets.

In pure IP networks, traffic engineering is implemented by changing the intradomain routing protocol (called Interior Gateway Protocol or IGP) weights [9]. Optimizing the IGP weights to force traffic to follow high-capacity and/or low-delay paths is a way to perform traffic engineering. When coupled with on-line TE algorithms, the label-based forwarding mechanisms such as Multi Protocol Label Switching (MPLS) [20] and Generalized Multi Protocol Label Switching (GMPLS) [14] make per-flow path selection with service guarantees possible. The label-based forwarding mechanisms provide an opportunity to control traffic in a finer-grained way than by changing IGP weights.

Traffic engineering requires the computation of paths between each pair of routers. This computation may be done off-line or on-line. The computation is done off-line when paths do not need

*Delft University of Technology, {B.Fu,S.P.W.G.Uhlig}@ewi.tudelft.nl

to adapt to traffic dynamics. On-line computation, on the other hand, is required when the routing plans must adapt to changing network conditions. On-line TE techniques leverage the capability of label-based forwarding mechanisms to choose paths inside the network to better utilize the available resources in the network. On-line TE algorithms select for each new connection request a path with enough resources, based on the current state of the network. In this paper, we focus exclusively on on-line traffic engineering algorithms.

On-line TE algorithms need to address two different issues:

- Find a routing path through the network with enough resources (e.g. bandwidth), called a *feasible path*;
- Try to accommodate as many requests as possible.

To use the network capacity in the most efficient way, the routing paths chosen for the traffic should not interfere with each other too much [12]. In practice, preventing interference between requests is difficult due to the low number of links in a network and the limited resources available.

Many factors affect the outcome of the path finding and resource allocation process by on-line TE algorithms. The topology structure, the link costs, the residual link capacities, the path selection, etc, affect the global performance of the TE algorithms. Obtaining a priori knowledge of the behavior of a TE algorithm on a particular network scenario is very hard due to the interactions between the different aspects of the problem.

TE algorithms are usually designed to optimize a particular objective function and address a specific traffic engineering problem [10, 12, 2, 24]. Setting up scenarios under which TE algorithms will be compared is difficult because some TE algorithms have not been designed to perform well under specific situations. Not all scenarios are practically relevant. If some TE algorithm performs poorly under a scenario that is not relevant in practice, this should not undermine the suitability of the TE algorithm.

In this paper we categorize the different factors that affect the behavior of TE algorithms as follows:

- network topology: The location of Points of Presence (POP) and the existing connectivity (e.g. fiber) between POPs largely determine the network structure. Most ISPs design their networks with specific redundancy requirements, leading to rather sparse topologies that trade-off cost and robustness [3].
- traffic demand: The amount of traffic between each pair of routers.
- network provisioning: Network provisioning consists in scaling the link capacities, so as to accommodate the traffic between all pairs of routers.

The network, traffic and routing are intertwined. The design of the network (capacities and link weights) must ensure that the traffic demand can be satisfied. To assign link capacities, network design requires an initial model of routing to determine how much traffic will flow on each link. In today's networks, link weights are set based on the delay, the capacity [18], or a combination of delay and capacity of the links. The goal of setting IGP weights in such a way is to attract traffic towards high-capacity and low-delay links. In practice, many backbone operators use the ad-hoc approach of observing the flow of traffic through the network [7], and iteratively adjusting a weight whenever the load on the corresponding link is higher or lower than desired. The problem has been addressed in [10, 9, 17] using different techniques from operations research. Observing traffic in large backbone networks requires significant resources [7], so that in practice the real traffic demand is inferred rather than observed [15, 25]. Few observed traffic demands are publicly available [24, 23].

In this paper, we study the behavior of on-line TE algorithms, using a set of real and synthetic traffic demands and backbone network topologies. To our knowledge, this is the first time that the relationship between network design and the behavior of on-line TE algorithms is studied in such a way. Our contributions consist in identifying several factors that determine the relevance of on-line TE algorithms. Due to the small viability of different time scales [22], in this paper we use arbitrary scalings, which are supposed to show the general behaviors.

The remainder of this paper is organized as follows. We review the literature by presenting some well-known TE algorithms in Section 2. The behavior of TE algorithms on two real-world scenarios is studied in Section 3. In Section 4, we build our own scenarios to study the impact of network design aspects on TE algorithms.

2. RELATED WORK

Many Autonomous Systems (ASes) use Open Shortest Path First (OSPF) [16] or Intermediate System-Intermediate System (IS-IS) [4] as their intradomain routing protocol. These protocols select shortest paths based on static link weights. These IGP link costs reflect the desirability of a link to be selected to carry traffic inside the AS. We call the routing algorithm used by these routing protocols the *Static Shortest Path* algorithm (SSP), as the shortest paths will remain the same no matter the amount of traffic actually carried on those links, as long as the link weights do not change. The link costs can be computed off-line according to the offered traffic, the provisioned network capacity, and the specific objective desired by the network administrator [10], or simply by following Cisco's guideline of IGP costs proportional to the inverse of the link capacity [5]. Recently, splitting traffic among several shortest-path, called equal-cost multi-path (ECMP) [16], has been deployed by an increasing number of network operators. ECMP provides some load-balancing and makes the granularity of the traffic finer compared to non-ECMP.

The algorithms that use the off-line computed link costs, are called off-line routing algorithms. SSP is computationally efficient, but unaware of the link utilization. If the link costs setting is far from the optimal setting with respect to the current traffic demand, then SSP may heavily load some links while avoiding others that have plenty of unused capacity. When capacity is not available on the shortest path, IP packets are simply dropped as the router buffers saturate or connection requests are rejected as feasible paths do not exist.

On-line routing algorithms have been proposed to find paths with bandwidth requirements. On-line algorithms use dynamic information such as the link residual capacities to compute the feasible paths, and are able to find one, provided it exists. There is an extensive literature on on-line TE algorithms. Due to space limitations, we do not discuss all of them. In this paper, we select a set of representative TE algorithms, from simple to computationally complex ones.

The Widest Shortest Path algorithm (WSP) [11] computes the shortest IGP paths in the network formed by links with sufficient residual capacities to accommodate the incoming request, and selects from all the feasible shortest IGP paths the one with the maximum bottleneck residual capacity to load-balance network traffic. It avoids using heavily loaded links unless there is no other option. Only feasible shortest IGP paths are considered.

Another instance of on-line algorithm is to use shortest-path routing with dynamic link weights. Instead of using SSP with a static inverse of the link capacity as the link weights, one can use the inverse of the residual capacity. As links get loaded, the weights will adapt to reflect the available capacity in the network. We call this technique dynamic shortest path with inverse residual capacity

(DSP-Inv).

The family of Minimum Interference Routing Algorithms (MIRA) [12] tries to select, between each source-destination pair, a path that interferes the least with other source-destination pairs. MIRA takes into account the information of source-destination pairs (S_i, D_i) and weights them by their importance α_i. The importance of a source-destination pair can be set for example as the fraction of the total traffic it represents. To minimize interference, those algorithms maximize the sum of the residual weighted max-flows[†] between all source-destination pairs. Upon the arrival of a connection request from S_i to D_i, the algorithm will select a path that can maximize

$$\sum_{(S_j,D_j),j\neq i} \alpha_j \theta_j \tag{1}$$

where θ_j represents the residual max-flow for (S_j, D_j) after a path is selected for (S_i, D_i). Maximizing this objective function is NP-hard [12]. The MIRA algorithms reach a suboptimal solution by computing the set of *critical* links L_j for all other pairs $(S_j, D_j)(j \neq i)$. A link l is called *critical* for a source-destination pair (S, D), if the reduction in l's capacity leads to the reduction in (S, D)'s residual max-flow. Then link costs are set according to the link *criticality*, and shortest path computation is applied to this weighted topology. When only the importance of each source-destination pair is considered in the link *criticality*, the link cost can be set as

$$C(l) = \sum_{(S_j,D_j),l\in L_j} \alpha_j \tag{2}$$

or as

$$C(l) = \sum_{(S_j,D_j),l\in L_j} \alpha_j / R(l) \tag{3}$$

when the residual capacity of each link is also considered, where R(l) is the residual capacity of l. When MIRA is using equation 2 (equation 3) for its link costs, it is called MIRA-TM (MIRA-TM-Cap). For further details about MIRA we refer to [12].

3. BEHAVIOR OF TE ALGORITHMS ON REAL-WORLD NETWORKS

In this section, we observe the behavior of on-line TE algorithms on real-world networks and real traffic demands. In Section 4 we rely on generated topologies to show the impact of different aspects of network design on the behavior of TE algorithms.

3.1. SCENARIOS

We use real topologies and traffic matrices from Abilene[‡] and GEANT[§]. Abilene and GEANT are the US and European academic backbones. They both provide transit service to universities and research institutions in the US and Europe. The Abilene topology has 12 nodes and 30 directed links while GEANT has 22 nodes and 72 directed links. The link capacities and link costs are kept as in reality, with link costs in Abilene assigned based on the geographical distance, and that in GEANT

[†]The max-flow of a source-destination pair (S, D) is the maximum amount of traffic that can be pushed between this pair. Multiple link-disjoint paths may be used.
[‡]`http://abilene.internet2.edu/`
[§]`http://www.geant.net/`

following a modified version of Cisco's proposal (proportional to the inverse of the link capacity). For Abilene [24], a traffic matrix spans 5 minutes, and for GEANT [23] 15 minutes. When using a traffic matrix, for each source-destination pair, we deduce the average bandwidth requirements from the traffic matrix. The traffic between each source-destination pair is further split into small pieces to create the requests for path establishment inside the network. The traffic demand from the traffic matrix between each source-destination pair is split into 200 equal pieces. In practice, the order in which the connection requests arrive may have an impact on the behavior of the TE algorithm. If the order of the connection requests is important for the performance of the TE algorithm, then an off-line TE algorithm should be used in order to properly schedule the requests. As we are interested in on-line TE algorithms in this paper, we do not consider the interactions between the connection requests. Therefore, the requests we use are small enough to be close to a fluid for the network¶. Connection requests are handled in the following way:

- Select uniformly an source-destination pair from all potential ones;
- Inject a connection request between this pair with a corresponding bandwidth requirement;
- Reserve the capacity for this connection request if a feasible path is found by the algorithm; the capacity will reserved for this connection until the end of each analysis.

No matter which algorithm is used to compute the paths (including SSP), a reservation is made in the network if a feasible path is found by this algorithm.

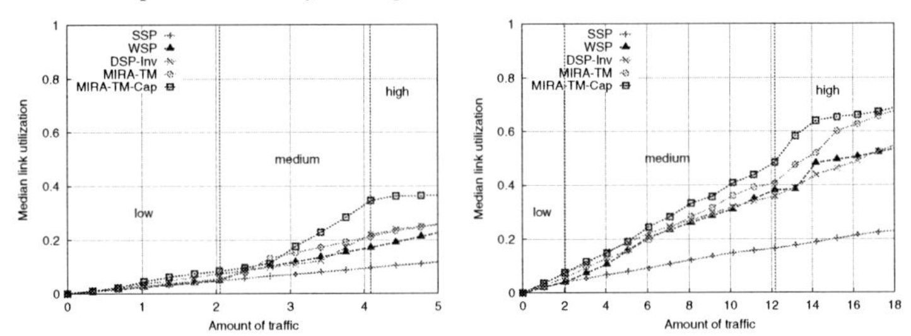

Figure 1: Median link utilization under different load regimes: Abilene (left) and GEANT (right).

By adjusting the total number of injected connection requests, different *load regimes* can be reached. We show in Section 3.2 how different *load regimes* appear when scaling traffic demands.

To check our results with the literature, we also borrowed the scenario from [12]. In this topology, each link is bidirectional (can be treated as 2 independent unidirectional links with the same capacity). Some links have capacities of 1200 units and the others have 4800 units. The connection requests arrive randomly, with the same rate for all given source-destination pairs, and with a bandwidth requirement uniformly distributed between 1 and 3 units. The connection requests are injected into the network in the same way as explained for Abilene and GEANT.

3.2. IMPACT OF TRAFFIC LOAD

In this section we provide simulation results of the different TE algorithms applied to both Abilene and GEANT. For each network, we pick one traffic matrix out of several measured ones

¶We considered coarser connection requests and noticed that increasing their size does not significantly change our results, but makes the curves appear less smooth.

during peak time. Figure 1 shows the median link utilization after each connection request is routed by the TE algorithms. The median link utilization is the value such that half of the links in the network have a smaller utilization. We use the median because we want to have a picture of the global link utilization in the network, and because it is less sensitive to extreme values than the average. The average link utilization provides similar results. The wide variations between the minimum and the maximum link utilization make those statistics not insightful.

To see how TE algorithms manage to handle different traffic loads, we inject requests proportionately to the real traffic matrices until the connection requests for a significant fraction of the source-destination pairs cannot be accepted. By doing this, we ensure that we sample all *load regimes* that correspond to the traffic distribution given by the traffic matrix.

The results for the following TE algorithms are shown on Figure 1: SSP, WSP, DSP-Inv, and 2 variants of MIRA (MIRA-TM and MIRA-TM-Cap). MIRA-TM considers in its weighting the traffic between source-destination pairs (see equation 2). MIRA-TM-Cap takes into account not only the traffic between source-destination pairs, but also the residual capacity of the critical links (see equation 3).

The x-axis of both graphs of Figure 1 give the sum of the connection requests that have been pushed into the network so far (total traffic that has been accepted). We scaled this sum to quantify the traffic demands in terms of the measured traffic matrices, without revealing the exact amount of traffic that a single traffic matrix represents. The left graph in Figure 1 makes three different *load regimes* appear: low load, medium load, and high load. When amount of traffic is low enough, the algorithms can be divided into two groups: 1) SSP and WSP; 2) DSP-Inv and the MIRA family. SSP and WSP share the same link costs settings which do not give multiple shortest IGP paths in the Abilene network, so when the load is low enough SSP and WSP will choose the same paths. DSP-Inv and the MIRA algorithms on the other hand do not follow the static link costs settings of the network, but use longer paths with respect to the static IGP costs to avoid critical links and hence reduce interference with other requests. If network administrators prefer low link utilization, TE algorithms like DSP-Inv or MIRA-TM-Cap perform worse than SSP and WSP as seen from the graph for median link utilization.

In what we call the medium load regime, SSP and WSP choose different paths. WSP chooses longer paths on average, like MIRA, hence the median link utilization increases faster than for SSP. When the network starts to become loaded, WSP detects that some shortest paths are not feasible any more and thus turns to alternative paths. SSP always uses the paths computed based on the static link costs settings, resulting in blocked connection requests when load becomes too high. One of the interests of TE algorithms lies in their ability to find feasible paths as long as there exists some, even when load is becoming high in the network.

Finally, the high load regime occurs when not enough capacity is left in the network. In this situation, even the MIRA family is unable to satisfy connection requests between a significant fraction of the source-destination pairs. This situation should never happen in practice, as it indicates inadequate network provisioning.

For the GEANT network (right graph Figure 1), we also observe three distinct load regimes. In the low load region SSP and WSP behave in the same way. Then in the medium load region WSP chooses different paths from those chosen by SSP and has a higher median link utilization. Finally, in the high load regime, the median link utilization hardly increases due to lack of available capacity to satisfy connection requests.

3.3. AVAILABLE CAPACITY IN THE NETWORK

Figure 1 showed that the median link utilization stays relatively low, even under the high load regime. Although many links still have capacity left, TE algorithms are unable to satisfy a significant fraction of the connection requests. As paths between source-destinations are typically made of several links, connection requests between most of the source-destination pairs cannot be satisfied, as soon as a significant fraction of the links are highly loaded.

Link utilization-related metrics give a picture that is easy to understand for network operators, as it is closely related to delay [19]. To understand the behavior of TE algorithms on the other hand, a metric based on the capacity left in the network is more helpful. For this, we rely on the sum of the residual max-flows of all source-destination pairs. The residual max-flow of a particular source-destination pair tells how much usable capacity is left in the network to satisfy connection requests between the considered source-destination pair. The sum of the residual max-flows for all source-destination pairs tells how much capacity is usable by TE algorithms to satisfy connection requests.

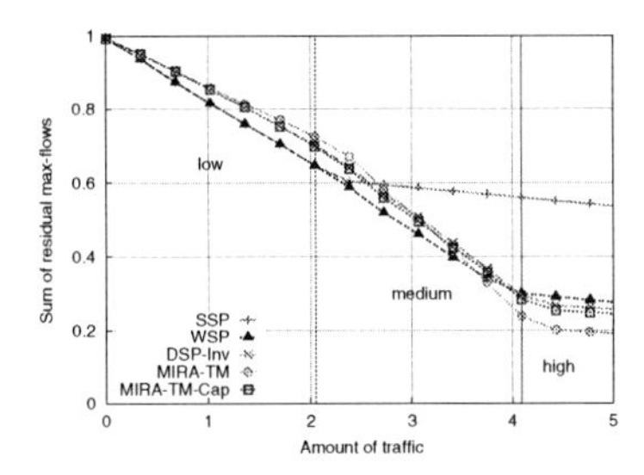

Figure 2: Network-wide residual max-flow on Abilene network.

Figure 2 shows how the sum of residual max-flows for all source-destination pairs evolves for increasing load in the Abilene network (x-axis is the same as on Figure 1). Note that the value of the sum of the residual max-flows is meaningless, only the relative speed at which it decreases for different TE algorithms is of relevance. We thus normalize it in all figures.

In the low load regime, there is no big difference between the algorithms. SSP and WSP use a bit more of the total max-flow than the other three algorithms. In the medium load regime, SSP cannot satisfy some connection requests, hence does not use the available max-flow. WSP, DSP-Inv and the two MIRA variants manage to explore available paths to route connection requests. In the high load regime, the four dynamic algorithms hardly manage to satisfy connection requests. Hence, they have a very slowly decreasing residual max-flow.

Overall, on-line TE algorithms behave in a similar way. They use longer paths than SSP to make a better use of the available capacity. Unless those longer paths do not interfere with later requests from other source-destination pairs, TE algorithms should perform equally well on the scale of the whole network, as measured by the sum of the residual max-flow. As networks contain less links than source-destination pairs, it is unlikely that alternative paths that do not interfere with other source-destination pairs may be found in the network.

To better understand why TE algorithms behave globally in a similar way, we need to focus on specific source-destination pairs. To make the explanation simpler, we take the same topology as used in [12]. The MIRA algorithms were proposed to try to minimize interference between different source-destination pairs. The insight behind MIRA is that in order to better utilize the available

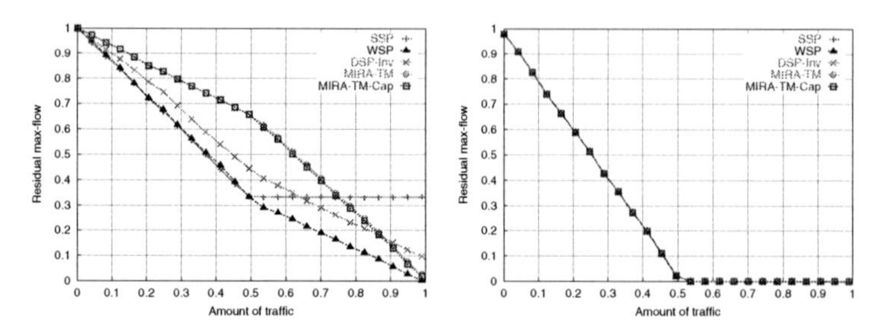

Figure 3: Residual max-flow for pair (S_1, D_1) (left) and pair (S_2, D_2) (right)

capacity in the network, each source-destination pair should try to prevent interfering with the residual max-flow of other source-destination pairs when choosing their path. Another important aspect of interference is that different source-destination pairs have different possibilities to have their paths routed in order to minimize interference. We implemented the scenario of [12] and computed after each request the residual max-flow of each source-destination pair.

Figure 3 shows the residual max-flows of 2 of the 4 source-destination pairs used in [12], (S_1, D_1) and (S_2, D_2). For (S_1, D_1), we obtain results similar to those shown in [12]. The two variants of the MIRA algorithm leave a larger residual max-flow compared to the other three algorithms for an amount of traffic smaller than 0.75. Next comes DSP-Inv, and finally WSP and SSP. At an amount of traffic of 0.5, SSP cannot use the whole max-flow of (S_1, D_1) due to its inability to choose alternative paths. DSP-Inv is also unable to use the whole max-flow of (S_1, D_1), although to a smaller extent than SSP. WSP and the two MIRA variants on the other hand manage to use the whole max-flow of (S_1, D_1).

For (S_2, D_2) (bottom of Figure 3), the residual max-flows found for all the considered algorithms are the same. Although there are two bottleneck links forming (S_2, D_2)'s max-flow, both links are also used by (S_4, D_4)'s max-flow. No TE algorithm can prevent interference to happen for (S_2, D_2).

From a network-wide perspective, TE algorithms alleviate problems on some links by shifting it to other links. How much TE helps, however, depends much on the considered source-destination pair. When alternative paths can be used without interfering with other source-destination pairs, TE algorithms may help. When critical links cannot be bypassed, TE cannot compensate for inadequate provisioning in the network or for links that belong to the residual max-flow of several source-destination pairs.

3.4. IMPACT OF TRAFFIC PATTERN

Network traffic is dynamic, it exhibits daily and weekly patterns [6, 21, 8]. Different traffic patterns might complexify the picture of TE algorithms we have shown so far. In the previous section, we showed the impact of the load regime on the behavior of TE algorithms. Over time, both the total amount of traffic and the distribution of the traffic among source-destination pairs may change. On Figure 4, we show the distribution of the traffic among source-destination pairs for four different traffic matrices of Abilene. We selected two days, June 1st and 2nd 2004, and within those two days we selected two time intervals, one during peak time (16:55) and another during non-peak time (21:35).

We observe on Figure 4 that the distribution of the traffic differs much between peak and non-peak time for Abilene. During peak time, one particular source-destination pair is responsible for about 60% of the total traffic. During non-peak time on the other hand, the traffic distribution among source-destination pairs is less uneven, but still far from uniform. During non-peak time, 20 among the 132 source-destination pairs are responsible for about 50% of the total traffic. Note that the variations in the traffic demand of GEANT are limited and do not impact our results.

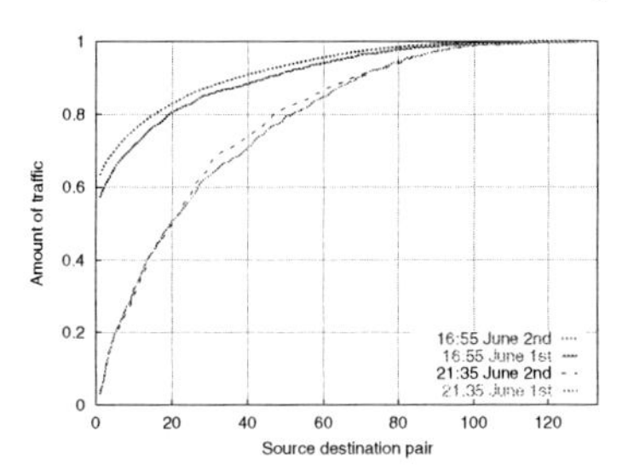

Figure 4: Distribution of traffic among source destination pairs (Abilene).

Similarly to Section 3.2 and 3.3 where the results under traffic matrix of Abilene taken at 16:55 June 1st were shown, we performed the comparison of TE algorithms on several other traffic matrices (the other three instances whose distributions are shown in Figure 4) of Abilene to see how different traffic matrices affect the comparison. Due to the space limitation, we do not show the results for these traffic matrices in this paper. Surprisingly, we do not observe major differences with the results from Section 3.2 and 3.3. As soon as the network becomes loaded, we observe different behaviors of the TE algorithms, corresponding to the medium or high load regimes. When the amount of traffic is low, then the TE algorithms behave as in the low load regime.

While traffic matrices for 16:55 June 1st and 16:55 June 2nd push the network to experience the 3 load regimes with traffic scale 5, the figures for 21:35 June 2nd show a linear increase in median link utilization and decrease in residual max-flow. The reason is that the traffic matrix taken at 21:35 corresponds to relatively low traffic, compared with the ones taken at 16:55 which correspond to peak time.

3.5. TE ALGORITHMS AND LOAD REGIMES

From the results of this section, we identified one main factor that affects the behavior of the TE algorithms: the total amount of traffic the network has to handle. This total amount of traffic translates, through the choice of the paths made by the routing algorithms, into a load regime. We are now in a position where we can loosely define the three load regimes we observed in this section. The low load regime is defined by a behavior of the TE algorithms that is similar to the one of SSP. In this situation, it is not necessary to use non-shortest paths to accept the connection requests. As soon as SSP is not able to satisfy some connection requests, we enter the "medium load regime". In this medium load regime, TE algorithms manage to better use the available capacity than SSP, to accept connection requests that would be blocked otherwise. Finally, when the amount of traffic is so high, that even TE algorithms are unable to satisfy connection requests, we enter the "high load regime". In this high load regime, the solution is to increase the capacity of the links or to add redundant links.

4. IMPORTANCE OF NETWORK DESIGN ASPECTS

Section 3 showed the differences between TE algorithms under different load regimes. As we relied on real-world networks and traffic demands, we could not pinpoint the importance of specific aspects of network design on the behavior TE algorithms. In this section, we build up scenarios to study the impact of the network topology and network dimensioning. We propose scenarios based on differently connected topologies, with different ways of assigning link capacities and generating traffic demands.

Topology Our topologies are generated using the iGen topology generator[‖]. iGen allows to generate random points in one or any continent, and then to connect the nodes using network design heuristics [3]. It can also set capacities of the links and IGP link weights (based on physical distance or the inverse of the link capacity). We choose topologies with 25 nodes, and randomly generated points in Northern-America.

Link weights The weight for each link l_{ij} is assigned as a piecewise linear function of the geographical distance between node i and j.

Traffic demand and capacity provisioning For each topology, two combinations of traffic demand and capacity provisioning are considered:

- *properly provisioned networks:* We generate the traffic demand based on a gravity model [13]. First, we generate for each edge node i (source or destination) a total traffic demand x_i following a uniform distribution. The amount of traffic X_{ij} from node i to node j is set proportional to the product of the traffic demands x_i and x_j, i.e., $X_{ij} = \beta x_i x_j$, where β is some constant. We assign capacities to links by first assigning paths to source-destination pairs. Paths are obtained by running a shortest path using the already assigned link weights. Once a link is used by the path from node i to j, the capacity of this link will be increased by 1.1 times the corresponding traffic amount X_{ij}. The factor 1.1 provides a very small over-provisioning. When links are not used by any shortest path, they would end up with 0 capacity. We assign to these links the average capacity of the non-empty links.

- *Improperly provisioned networks:* The traffic and link capacities are considered separately. All links are assigned the same capacity. All source-destination pairs have the same total amount of traffic, but connection requests have a uniform size between 1 and 3 units of traffic.

In the *properly provisioned networks*, the capacities of links are assigned in such a way that the provisioning of the network matches the traffic matrix, under the assumption that shortest paths are used. In *improperly provisioned networks*, the network capacities are not designed to match the traffic demand at all. We thus expect different behaviors of the TE algorithms under the two types of networks.

4.1. MINIMALLY-CONNECTED TOPOLOGY

We start by relying on the worst possible connectivity a network can have: a tree. We use the MST (minimum spanning tree) heuristic from iGen to generate the topology. There are 24 links in this topology. All the algorithms perform in the same way (figures not shown), whether or not the network

[‖] http://www.info.ucl.ac.be/~bqu/igen/

is properly provisioned. For any source-destination pair, only one path is physically available, so all algorithms have no choice but to choose the same path.

4.2. WELL-CONNECTED TOPOLOGY

Real-world topologies are typically designed to have minimal cost, while being able to stand any link failure. Several methods exist to generate such graphs [3]. We use the Two-Trees heuristic from iGen to generate a graph made of two link-disjoint MSTs. Such a graph remains connected after any single link failure. At least 2 paths exist between any pair of nodes. The topology has 48 links.

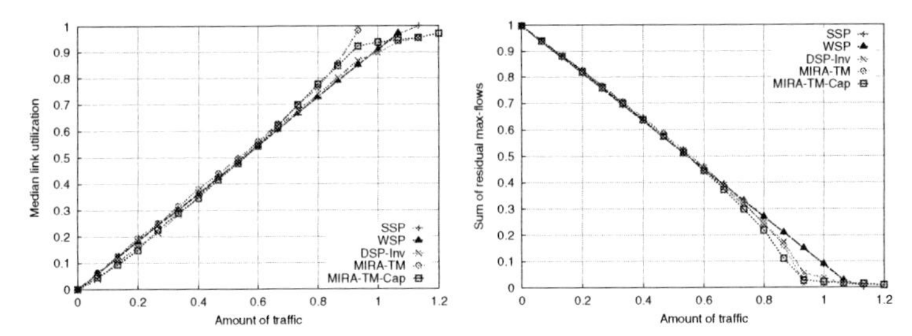

Figure 5: Median link utilization (left) and network-wide residual max-flow (right) in properly provisioned Two-Trees topology.

4.2.1. PROPERLY PROVISIONED NETWORK

Figure 5 shows the median link utilization and the residual max-flow after each connection request on the properly provisioned Two-Trees topology. There is not much difference between the TE algorithms, both in terms of the median link utilization and the sum of the residual max-flows. As the network provisioning matches the traffic matrix, all algorithms manage to reach a high median link utilization (0.93) when the demand equals the network capacity (amount of traffic = 1). It is interesting to note that both SSP and WSP have a linear increase (resp. decrease) of the median link utilization (resp. sum of residual max-flow) until the network is fully loaded. As SSP and WSP use shortest paths that better match to the way the network was provisioned, their behavior is more appropriate in this network than more complex TE algorithms.

4.2.2. IMPROPERLY PROVISIONED NETWORK

Figure 6 shows the median link utilization and the residual max-flow after each connection request on the improperly provisioned Two-Trees topology. In well-provisioned networks, TE algorithms are hardly useful. In networks that have not been provisioned in such a way as to match the traffic demand, the typical low-medium-high load regimes appear quite early, when the median utilization is rather low.

In the low load regime, all algorithms give the same value of the sum of residual max flow (bottom of Figure 6). In the medium load regime, SSP blocks connection requests while all other algorithms manage to find feasible paths in the network.

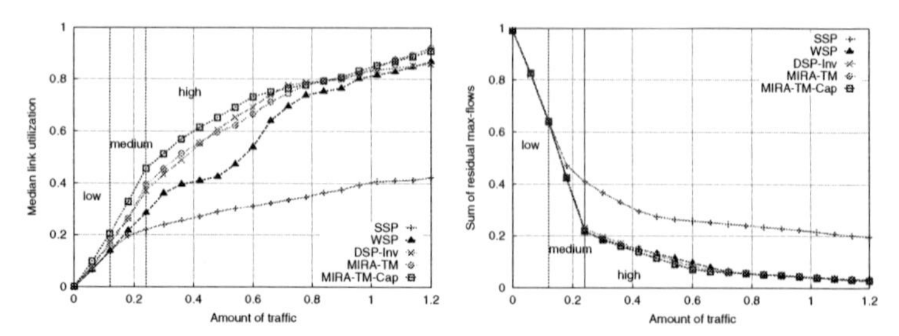

Figure 6: Median link utilization (left) and network-wide residual max-flow (right) in improperly provisioned Two-Trees topology.

4.3. HIGHLY-CONNECTED TOPOLOGY

If local redundancy is required in a topology, any three nodes close to each other can be connected by a triangle. The Delaunay triangulation procedure ensures such a connectivity, leading to locally well-connected graphs. Using the Delaunay heuristic in iGen to connect the 25 nodes, we obtain a topology with 65 links.

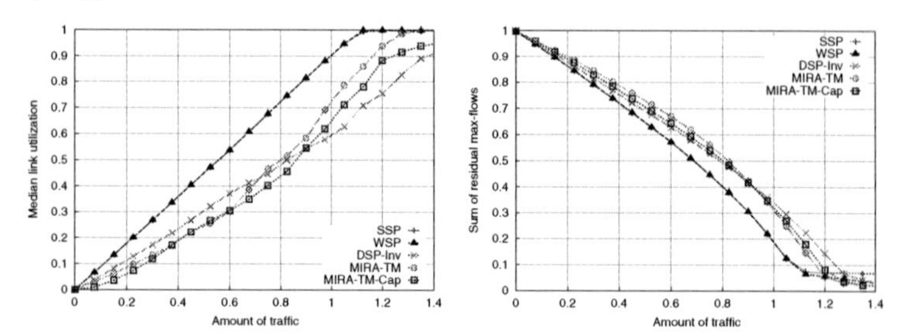

Figure 7: Median link utilization (left) and network-wide residual max-flow (right) in properly provisioned Delaunay topology.

4.3.1. PROPERLY PROVISIONED NETWORK

As this topology provides much redundancy, when assigning link capacities in the properly provisioned scenario, some links that are never used by the shortest paths are assigned a capacity within the same scale as other links (their average). As shown in Figure 7, SSP and WSP have a median link utilization that increases linearly with the amount of traffic until it reaches a value of 1, faster than the other algorithms. DSP-Inv and the MIRA algorithms manage to leave a higher residual max-flow than SSP and WSP. The good provisioning and link redundancy allows all algorithms to use available capacity even when the link utilization becomes high. Contrary to the well-provisioned Two Trees topology, the redundancy of the Delaunay topology allows algorithms to choose very different paths, hence the differences between the algorithms.

4.3.2. IMPROPERLY PROVISIONED NETWORK

Figures 8 shows the median link utilization and network-wide residual max-flow after each connection request on the improperly provisioned Delaunay topology. In the low load regime, DSP-Inv and the two MIRA variants manage to route the requests while leaving a larger residual max-flow than SSP and WSP. SSP keeps a low link utilization at the cost of blocking connection requests, while WSP has the highest median link utilization not to block requests. DSP-Inv and the two MIRA variants manage to both keep a relatively low median link utilization while leaving a residual max-flow larger than WSP.

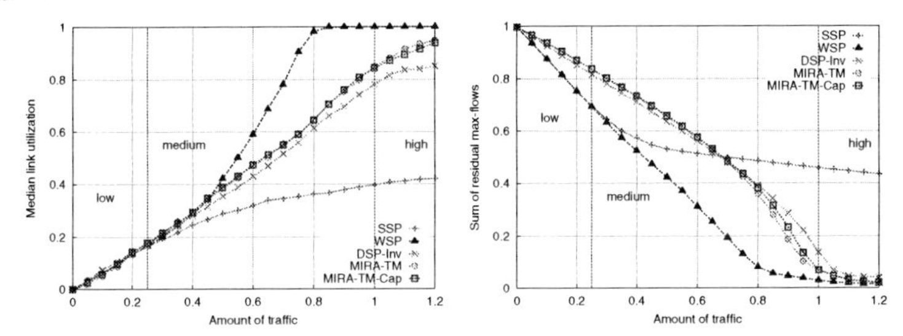

Figure 8: Median link utilization (top) and network-wide residual max-flow (bottom) in the improperly provisioned Delaunay topology.

4.4. DISCUSSION

In Section 3 we studied the behavior of TE algorithms on two real-world networks: Abilene and GEANT. Those two networks have different topological structure, as well as different strategies to set up their IGP weights. The Abilene network is very close to a Minimum Spanning Tree, with 15 undirected links for 12 nodes. The Abilene network has been designed to have minimal number of links, while still being 2-connected, i.e. any single link failure still leaves the network connected. Abilene nodes have node degree 2 or 3. The GEANT network on the other hand is closer to a Two Trees topology than an MST, with 36 undirected links for 22 nodes. GEANT nodes have node degree from 2 to 8, so this network has been designed with far more redundancy than Abilene. Both Abilene and GEANT are over-provisioned networks, the link utilization is pretty low. Over-provisioned is not equivalent to what we call "properly provisioned" in this paper. Proper provisioning means that the capacity of the links in the network match the traffic between all source-destination pairs. In Section 3, we observe that when pushing these networks to their limit, they behave like improperly provisioned networks. It is unlikely that these networks relied on capacity dimensioning to match the traffic demand. Instead, they probably use a simple over-provisioning strategy.

5. CONCLUSION

In this paper, we studied the behavior of on-line TE algorithms under different scenarios. We first relied on two real-world networks, Abilene and GEANT, and their traffic. By scaling their traffic demand, we observed the behavior of TE algorithms when networks are pushed to their limits. We identified three distinct load regimes (low, medium, and high) that correspond to different behaviors of the TE algorithms. In the low load regime, TE algorithms do not provide much benefit compared to shortest-path routing. In the medium load regime, where shortest-path routing blocks some connection requests, TE algorithms manage to better use the available capacity in the network. Finally, the high load regime corresponds to a situation where not enough capacity is left in the network, so TE algorithms cannot route connection requests. In high load regime, network capacity or redundant links should be added.

In the second part of the paper, we studied the behavior of TE algorithms using synthetic network topologies and traffic demands. We compared the behavior of each type of topology under two network provisioning scenarios. In the first scenario, we provisioned the link capacities as to match the traffic demand. In the second scenario, we provisioned the network without taking into account the traffic demand. As long as the network is properly provisioned, TE algorithms do not provide a significant improvement in using the available capacity, even in highly-connected topologies where many paths are available. When network provisioning does not match the traffic demand, TE algorithms are able to use the redundant links to compensate for the poor provisioning.

We expect that most real-world networks are in a low load regime, where shortest-path routing is good enough. However, as the traffic demand evolves, networks need be updated. One potential further work is to develop a metric that quantifies how badly the current network does not match the current traffic demand. This metric would take into account the network growth and the routing algorithm used, and give insight into when the situation is becoming critical enough so that either TE algorithms should be used, or when the network must be upgraded. We also expect that the topological structure has a non-trivial impact on the behavior of the TE algorithms.

6. ACKNOWLEDGMENTS

Bingjie Fu is supported by the Dutch Technology Foundation STW under the project number DTC.6421. We thank Abilene and GEANT for providing their topology and traffic demands. We are also grateful to Bruno Quoitin, Almerima Jamakovic and Piet Van Mieghem for comments on previous versions of this paper.

REFERENCES

[1] AWDUCHE, D., CHIU, A., ELWALID, A., WIDJAJA, I., AND XIAO, X. Overview and principles of internet traffic engineering. Internet Engineering Task Force, RFC3272, May 2002.

[2] BLANCHY, F., MÉLON, L., AND LEDUC, G. An efficient decentralized on-line traffic engineering algorithm for MPLS networks. In *Proc. of ITC-18, Berlin, Germany* (2003).

[3] CAHN, R. *Wide area network design: concepts and tools for optimization*. Morgan Kaufmann Publishers Inc., San Francisco, CA, USA, 1998.

[4] CALLON, R. Use of OSI IS-IS for routing in TCP/IP and dual environments. *Request for Comments 1195, IETF* (December 1990).

[5] CISCO. OSPF design guide. `http://www.cisco.com/warp/public/104/1.html`.

[6] CLAFFY, K., BRAUN, H., AND POLYZOS, G. Traffic characteristics of the T1 NSFNET backbone. In *INFOCOM* (1993).

[7] FELDMANN, A., GREENBERG, A., LUND, C., REINGOLD, N., REXFORD, J., AND TRUE, F. Deriving traffic demands for operational IP networks: methodology and experience. In *Proc. of ACM SIGCOMM, Stockholm, Sweden* (2000).

[8] FOMENKOV, M., KEYS, K., MOORE, D., AND CLAFFY, K. Longitudinal study of internet traffic in 1998-2003. In *Proceedings of WISICT'04* (2004).

[9] FORTZ, B., REXFORD, J., AND THORUP, M. Traffic engineering with traditional IP routing protocols. *IEEE Communications Magazine* (October 2002).

[10] FORTZ, B., AND THORUP, M. Internet traffic engineering by optimizing OSPF weights. In *Proc. of IEEE INFOCOM* (Tel-Aviv, Israel, March 2000).

[11] GUERIN, R., WILLIAMS, D., AND ORDA, A. QoS routing mechanisms and OSPF extensions. In *Proc. of IEEE Globecom* (1997).

[12] KODIALAM, M. S., AND LAKSHMAN, T. V. Minimum interference routing with applications to MPLS traffic engineering. In *Proc. of IEEE INFOCOM* (Tel-Aviv, Israel, March 2000).

[13] KOWALSKI, J., AND WARFIELD, B. Modeling traffic demand between nodes in a telecommunications network. In *Proceedings of ATNAC'95* (1995).

[14] MANNIE, E. Generalized multi-protocol label switching (GMPLS) architecture. *Request for Comments 3945, IETF* (October 2004).

[15] MEDINA, A., TAFT, N., SALAMATIAN, K., BHATTACHARYYA, S., AND DIOT, C. Traffic matrix estimation: existing techniques and new directions. In *Proc. of ACM SIGCOMM* (2002), pp. 161–174.

[16] MOY, J. OSPF version 2. *Request for Comments 2328, IETF* (April 1998).

[17] NUCCI, A., BHATTACHARYYA, S., TAFT, N., AND DIOT, C. IGP link weight assignment for operational tier-1 backbones. *IEEE/ACM Trans. Netw. 15*, 4 (2007), 789–802.

[18] ORAN, D. OSI IS-IS intradomain routing protocol. *Request for Comments 1142, IETF* (February 1990).

[19] PAPAGIANNAKI, K., MOON, S., FRALEIGH, C., THIRAN, P., TOBAGI, F., AND DIOT, C. Analysis of measured single-hop delay from an operational backbone network. In *Proc. of IEEE INFOCOM* (June 2002).

[20] ROSEN, E., VISWANATHAN, A., AND CALLON, R. Multiprotocol label switching architecture. *Request for Comments 3031, IETF* (January 2001).

[21] THOMPSON, K., MILLER, G., AND WILDER, R. Wide-Area Internet traffic patterns and characteristics. *IEEE Network magazine 11*, 6 (November/December 1997).

[22] UHLIG, S. Nonstationarity and high-order scaling in TCP flow arrivals: a methodological analysis. *ACM SIGCOMM Computer Communication Review 34*, 2 (April 2004).

[23] UHLIG, S., QUOITIN, B., LEPROPRE, J., AND BALON, S. Providing public intradomain traffic matrices to the research community. *ACM SIGCOMM Comput. Commun. Rev. 36*, 1 (2006), 83–86.

[24] WANG, H., XIE, H., QIU, L., YANG, Y., ZHANG, Y., AND GREENBERG, A. COPE: traffic engineering in dynamic networks. In *Proc. of ACM SIGCOMM'06, Pisa, Italy* (2006).

[25] ZHANG, Y., ROUGHAN, M., LUND, C., AND DONOHO, D. An information-theoretic approach to traffic matrix estimation. In *Proc. of ACM SIGCOMM* (2003), pp. 301–312.

Voice Communication, Computer Games, Talkspurts, Silence Periods

Gabor Papp and Chris GauthierDickey *

CHARACTERIZING AND MODELING MULTIPARTY VOICE COMMUNICATION FOR MULTIPLAYER GAMES

Over the last few years, the number of game players using voice communication to talk to each other while playing games has increased dramatically. Unlike traditional voice-over-IP technology, where most conversations are between two people, voice communication in games often has 5 or more people talking together as they play. We present the first measurement study on the characteristics of multiparty voice communications and develop a model of the observed talking and silence periods that can be used for future research, simulation, network engineering, and game development. Over a 3 month period, we measured over 7,000 sessions on an active multi-party voice communication server to quantify the characteristics of communication generated by game players, including group sizes, packet distributions, user and session frequencies, and speaking (and silence) durations. Unlike prior results, our measurements and models demonstrate that the speaking and silence periods fit a Weibull distribution.

1. INTRODUCTION

Multiparty voice communication (MVC) is an important application that needs to be studied and researched. In multiplayer computer games, for example, thousands of players can interact and communicate with each other using built-in voice chat systems (e.g., World of Warcraft[†]). Game consoles have added multiplayer lobbies for player interaction as selling features of the hardware. Further, multiparty voice communication is widely used for conference calls and voice chat software and clearly will be part of future online collaborative systems. Thus, research in this area benefits the future design of MVC systems, is relevant and interesting to online game development and deployment, and is important to ISP network engineers supporting and hosting online game and voice systems.

Our research is the first work to look at characterizing multiparty voice communications over the Internet, and in particular when it is used with multiplayer games. Previous voice-over-IP (VoIP) measurement work has looked at quality of service parameters such as packet loss, packet reordering and its effects on sound quality, or at network characteristics and support of VoIP between two parties. Instead, our research attempts to characterize the traffic, packet arrival rates, group sizes, session frequencies and durations, and speaking and silence periods in order to develop mathematical models for multiparty communication that can be used for simulation and modeling.

*Department of Computer Science, University of Denver, {pgabor|chrisg}@cs.du.edu
[†] http://www.worldofwarcraft.com/

In the past, multiparty voice communication was limited to small groups because player bandwidth was limited to modem speeds. However, new computer games often include voice communication as an essential part of the playing experience. A typical multiplayer game might have up to 40 players coordinating with voice communication while spelunking. This differs significantly from a Skype[‡] session between two people.

To conduct our measurement study, we set up a TeamSpeak[§] server which allows clients to join, set up channels, and communicate with other clients on the same channel. We then recorded traffic over a three month period and analyzed the resulting data.

Our results show that multiparty voice communication differs from traditional two-party communication. We see that overall traffic follows a sinusoidal curve with a peak around 8PM and a period of 24 hours. Group sizes tend to average 5 people while groups with more than 20 people are the least common. Further, silence and talking periods followed a Weibull distribution, which differs from prior research on voice communications.

The main contribution of this work is a characterization of traffic patterns, group patterns, and voice patterns through measurements. In particular, we model the talking patterns mathematically, based on the measured multiparty voice communication sessions. Further, the characterization of voice patterns has typically been done on small sets of data; our study measures patterns from thousands of hours of voice data with thousands of unique sessions. Thus, future research in MVC systems will be able to use our models to drive experimental simulations, game developers can use these models to understand the impact of adding voice communications on network traffic already generated by their games, and ISPs can use this information for provisioning servers for hosting MVC systems.

2. BACKGROUND

Voice patterns consist of *on* and *off* periods (also called talkspurts and silence periods). Over the last 40 years, research has looked at these patterns and shown that the *on* periods follow an exponential distribution [3, 15] in traditional telephony. These results are important because they allow designers of hardware, codecs, and network administrators to predict the patterns of speech with mathematical models. Our research follows in this tradition, though we look at multiparty voice communication and study several orders of magnitude more sessions than previous research.

Markopoulou et al. measured the quality of voice communications over the Internet. They measured delay and loss over wide-area backbone networks and used these results with a voice quality model [10] to determine the efficacy of VoIP over the Internet for voice communication. The authors show that while many Internet backbones have sufficiently low delay, delay variability, and loss, several provide poor VoIP quality. Our work measures traffic directly at the server since we cannot provide a client that generates end-to-end measurements. However, we are more interested in the actual traffic patterns generated by the multi-party communications and less interested in whether or not Internet backbones support VoIP.

Jiang et al. looked at the on-off patterns in VoIP by recording and digitizing conversations and then applying *gap* detectors to determine how long people talked and how long they were silent [8]. Their results show that the length that people talk for somewhat follows an exponential distribution while the gap they are silent for deviates significantly from the same distribution. Our measurements differ and show that the on-off patterns of VoIP in multiparty communications follow Weibull distri-

[‡]`http://www.skype.com/`
[§]`http://www.goteamspeak.com/`

butions more accurately, without significant deviation.

Skype [1], a VoIP application, was measured by Chen et al. in order to determine the level of user satisfaction [4]. By measuring network traffic characteristics, they correlated the amount of jitter and interactivity of a session with the length of the call. Our work measures similar traffic characteristics, but measures data between multiple parties. However, we do not examine or predict conversation quality based on traffic characteristics.

Borella analyzed game traffic from a popular online game server on a LAN and modeled the inter-packet arrival time and packet sizes with extreme distributions [2]. Their model was validated using the λ^2 test, which we also use to validate our models. The λ^2 test is important in our situation because we have large data sets with over 180k sample points and χ^2 tests perform poorly in these situations [2].

Henderson and Bhatti modeled network traffic of an online game over the Internet [7]. This work was measured over the Internet instead of over a LAN, providing a more realistic model. Their work shows a daily and weekly traffic pattern similar to what we measured with voice traffic: evenings have peak traffic while early mornings have the lowest traffic. Further, more traffic is seen overall on the weekends. Pittman et al. and Svoboda et al. had similar diurnal patterns in their measurement work on large-scale multiplayer games [14, 16]. In addition, other researchers modeled game traffic (packet sizes, arrival times, sessions) with similar results [5, 17].

Note that our early results were published in [11], but only included the initial CDFs of talkspurt and silence periods. However, in this paper, we present our full range of results and model multiparty voice communications.

3. TRACE COLLECTION

In this section, we describe the architecture of our network, the content of the server log file, the collection of the VoIP sessions and the procedure that we used to clean the collected data.

3.1. NETWORK SETUP

TeamSpeak is a group voice communication server that allows multiple people to connect using a TeamSpeak client, join *channels*, and talk simultaneously with other people in the same channel. In this client/server architecture, the clients encapsulate voice packets using one of many codecs, and send those packets to the server using UDP unicast. The server then unicasts the packets to the other $n-1$ clients connected on the same channel. Note that the server does *not* multiplex the voice packets, though we expect future architectures to do so.

We set up a TeamSpeak server and advertised it to game players as a free server beginning in November of 2006. We then began logging all traffic on port 8767 to the server using *tcpdump*. The server was set up to only allow 12.3kbps and 16.3kbps Speex encoding for voice.

Although TeamSpeak generates a server log file, the data contained in this file (even with maximum verbosity) is minimal and contains data only about logins, logouts, channel switching, and administrative operations. Thus, we used *tcpdump* to record all packet information generated with regards to the TeamSpeak server. We also discovered that when we compared data from the server log and the trace files, the server log was not always accurate. For example, the server log would show a player logging in multiple times without logging out. This was probably due to the fact that the player's connection died, but the server had not discovered it before the player re-logged in. However, from our packet traces we could determine a *session* by looking at the time that a player logged into the server to the last time they sent or received a voice packet from any player.

Using the data recorded by our tcpdump logs, we could identify voice packets and separate them from non-voice packets by their size and the codec ID. All voice packets were 155, 161, 205, and 211 bytes. Incoming packets were 155 and 205 bytes for 12.3 and 16.3 kbps Speex encoding. Outgoing packets were 161 and 211 bytes.

3.2. DATA CLEANING

As we began the analysis of our data, we discovered that some of the data points were extremely different from the rest. These included excessively long talk sessions as well as silent periods. For example, the extreme outliers of the talkspurts were data points where voice packets were delivered for close to an hour, which would be fairly difficult to accomplish when you consider that we can detect silence gaps as small as 100ms! We reckon that these are rare occasions resulting from something such as loud background music or human behavior such as forgetting to log off while leaving the computer for several hours. Therefore, we treated these data points as outliers and did not include them in our final results.

While removing extreme outliers can be controversial, we justify our actions by noting that our method removed few or no data points and that the methods used for curve fitting often pick the first and last end-points of the data to begin and end the curve, and then adjust values to force the rest of the graph to fit. Thus, the extreme outliers can cause a curve to not fit the data well, whereas by removing the outliers and fitting the curve allows one to obtain a better fit, according to various metrics. In Section 4.5, we detail the effects of our data cleaning.

To remove the extreme outliers, we first analyzed the linearity of the data. Prior researches show that talkspurt and silence periods often follow an exponential distribution [3, 8, 15]. We also plotted preliminary graphs to get an idea of the general trend of the data. Note that linearization for the purpose of cleaning does not need to be perfect (e.g., we used an exponential form, though our data turned out to be Weibull). The purpose of this process is to remove extreme outliers and, given the large number of data points, removing only a small percentage of data points is acceptable.

Once the data was linearized, we identified the first and third quartiles. To keep as much data as possible, we deleted only the extreme outliers. The data points, e, that are beyond the outer fences are defined as:

$$e < Q_1 - 3IQR \quad \text{or} \quad e > Q_3 + 3IQR \tag{1}$$

Here, Q_1 is the first quartile, Q_3 is the third quartile and IQR means the inter-quartile range $(Q_3 - Q_1)$. While methods that remove all the outliers and not just the extreme ones use $1.5IQR$ we decided to use $3IQR$ and only remove a very small amount of data, which we felt was sufficient enough for our purposes.

4. MEASUREMENTS

Our measurements cover a 3 month period from December 2006 to February 2007. During this time, we measured over 7000 sessions from over 800 IP addresses dispersed geographically for an average of 1.46 GB/day in traffic.

Table 1: Heaviest user distributions with rankings.

(a) Distribution by country.

Country	Player Distribution %
United States	60.53
Canada	26.63
Singapore	9.68
Australia	1.57

(b) Distribution by state.

State	Player Distribution %
Pennsylvania	17.80
New York	8.40
California	7.20
Colorado	4.60
Florida	4.20
Texas	3.80
Washington	3.20

4.1. GEOGRAPHICAL DISTRIBUTION OF USERS

In order to ensure that our data was not biased due to the geographical location of clients connecting to the server, we took advantage of the fact that all the client IP addresses were obtainable from our log files. Thus, we could estimate the locations of the clients and ensure that they were not all from the same place. Using the free MaxMind tool, GeoLite Country¶, we determined the latitude, longitude, country and state where applicable of each IP address. This free version of the software claims to have over 98% accuracy. After processing our data we found that more than 87% of our users were from North America, more than 60% of our users were from the United States, each of the 7 most popular states was responsible for more than 3% of the U.S. users and combined they were responsible for almost 50% of the U.S. users, and none of the remaining states contributed more than 3% of the population individually.

Table 1a shows that the majority of our users are from the United States and Canada while Table 1b shows the breakdown by states in the United States. Note that only two countries contributed more than 10% of the population, the U.S. and Canada. *We conclude that the primary result of our server location being in the MST time-zone is simply that most users are from the U.S. and Canada.* Generally, server location affects the user locations due to latency issues, but given the wide-spread locations of users within the continental U.S. and Canada, our data is not biased towards a particular area within these two countries, except to follow natural populations.

4.2. OVERALL SERVER TRAFFIC

The first set of measurements we present are the overall traffic seen by the server during an average day. Figure 1 shows the averages, averaged per hour on the x-axis and the number of packets sent and received on the y-axis. Thus, this figure is an indication of the volume of traffic seen by the TeamSpeak server. Incoming voice packets are always 155 or 205 bytes while outgoing voice packets are always 161 or 211 bytes respectively.

This result shows that server input doubled and server output increased by an approximate factor of 4 during the evenings (approximately 7pm-9pm MST‖). This indicates that more users are online using multiparty voice communication during the evenings.

¶`http://www.maxmind.com/app/geolitecountry`

‖Throughout this paper, times are listed as MST, but this is only as a convenience indicating the time-zone the server is located in and has no bearing on the measurements or results.

We also observed that during the peak period (7pm-9pm) the traffic rate is also almost constant. In other words, the number of sessions started is the same as the number of sessions finished during this period and thus resembles a balanced birth-death process.

We hypothesized that traffic was actually higher on weekends, and therefore we divided our averages into individual days so that we averaged all Mondays separately, all Tuesdays separately, etc. Figure 2 shows the inbound server traffic from users. From this figure, we see that most days are very similar with a small amount of variance, though in terms of total input, Fridays and Sundays have the highest amount of inbound traffic.

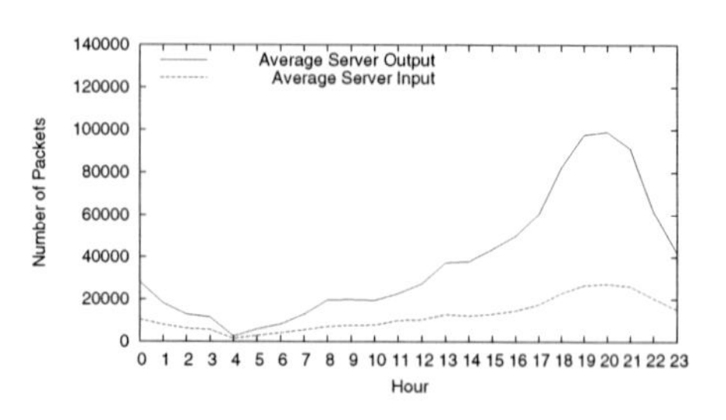

Figure 1: *Server Traffic:* Average voice server traffic over a 24 hour period (times shown are MST). Server input doubled while server output quadrupled during evening hours. The peak is around 7pm-9pm whereas the most quite period is around 4am.

In Figure 3, we can more clearly see that the server output has more traffic on weekends than on weekdays. In addition, Sunday traffic increases earlier than on any other day, starting at 1pm MST while the peak of traffic is highest late on Friday evenings at approximately 130k packets/hour. Interestingly, Saturdays have a lower peak traffic than Sundays or Fridays, but have a higher average traffic during the early hours of the day. This difference is most likely due to people who are on late Friday continuing to use TeamSpeak into the early hours of Saturday morning (and then probably sleep late that day).

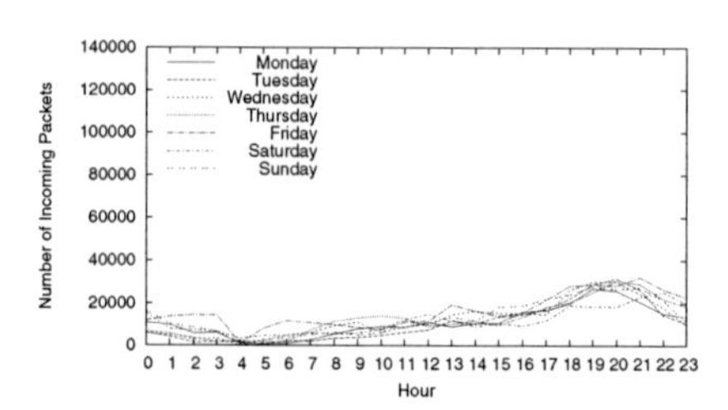

Figure 2: *Server Input:* Average voice server input traffic over a 24 hour period (times shown are MST). Server input is similar on all days, with a peak during evening hours.

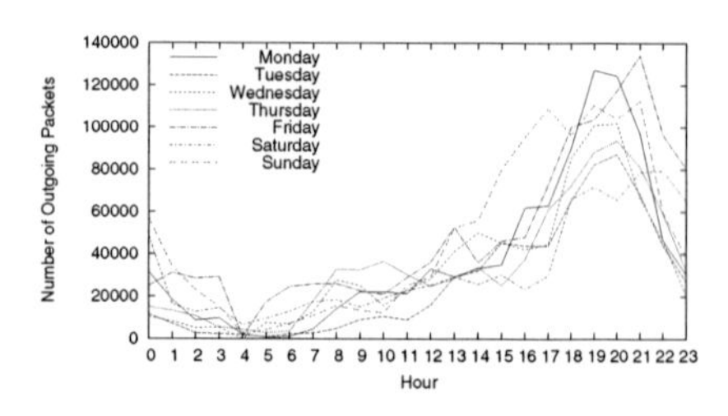

Figure 3: *Server Output:* Average voice server output traffic over a 24 hour period (times shown are MST). Server output is higher and extends over more hours during the weekends, and in particular on Sundays.

A final interesting trend is that Monday also seems to have a high outbound traffic peak that is similar to Friday. The primary difference appears to be that the traffic is shifted about two hours earlier, probably because game players start earlier so that they can go to bed earlier.

Given the TeamSpeak architecture, which unicasts packets to other players in the same channel, these results provide an insight into the size of the group that is talking to each other in the same channel. First, on Fridays, the output is approximately 4 times the size of the input. This implies that for each voice packet that is input, TeamSpeak is replicating it 4 times, for a group size of 5. On a day such as Tuesday, the traffic is 2 to 3 times that of the input, indicating group sizes on average of 3 to 4. We conjecture that on Fridays and Sundays, game players are more likely to use multiparty communication to converse with a larger group of other people than on other days. Most likely this is because players have more free time on those days and are able to coordinate getting together online with other players more readily.

When we look at this data in conjunction with the general server traffic, we see an interesting trend. Even though group sizes may increase, the amount of incoming traffic does not increase at the same rate as the outgoing traffic. Given these traffic patterns we believe that while many people may be able to talk at the same time in a large group, human protocols prevent this from occurring. Typically, only one person can talk at a time and they take turns during the sessions. In essence, if more than one person begins talking, the speakers stop to allow only one person to talk so that the conversation can be understood.

We expect our results to be similar to traffic patterns seen on game servers. Indeed, similar diurnal patterns and weekly patterns have been observed in related game traffic measurement work [17, 7, 14, 16]. There is clearly a peak, a local minimum each day, a strong correlation between days and a higher load on the weekends.

4.3. GROUP SIZES

We next examine group sizes to gain an insight into the size of a group that is typical in multiparty voice communication when used with games. As we noted previously, the ratio between the inbound and outbound traffic is an indicator of the average group size.

To perform this measurement, we looked at the trace logs and determined the sender ID for all the outgoing packets. The number of outgoing packets with the same ID and the same sequence number is one less than the actual group size. We binned all data according to how many people it was duplicated to, allowing us to examine the data based on the size of the group. Thus, we can determine the effect of the groups with different sizes on both the incoming and outgoing traffic on the server. Note that, neither the server log file nor the TeamSpeak packet format provides information about the used channel and thus it is impossible to identify the actual groups based on the packet alone. The results are seen in Figure 4. The solid line indicates the incoming traffic and the dashed line indicates the outgoing traffic that is simply the incoming traffic multiplied by the number of players that the particular packet is replicated.

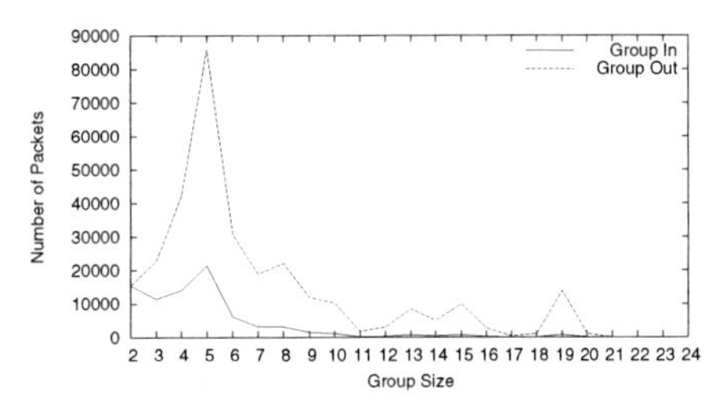

Figure 4: *Group Sizes:* The number of groups of a given size, categorized by days over the measurement period. We see that, on average, the most groups are between 2 to 8 people talking, with a maximum of 24 people in a group.

The results on group sizes show that the most active groups are the ones that are formed by 5 people. It can also be concluded that these groups generate the most outgoing traffic among all the groups. It is also worth mentioning that the amount of incoming packets from groups formed by pairs is almost as high as it is by these groups, but because of the replication the outgoing traffic is much less affected. Similarly, the counter effect can be seen in case of groups that contain 19 people;

although the amount of incoming traffic is low, the amount of outgoing traffic is high due to the large amount of replication necessary. The largest group we observed was 24 people.

We believe that a correlation between using TeamSpeak and the game being played exists. Currently, one of the most popular online games being played is World of Warcraft. In this game, players are often limited to 5 people in special areas, biasing the data towards a small group of people talking and playing the game together. On the other hand, a large class of multiplayer games, called *first-person shooters*, tend to group players into two groups, each between 8 and 16 players. Multi-party voice communication has also become very important for this class of games. If our TeamSpeak server was used by players of these kinds of games, we would expect the group sizes to correlate. Thus, we concluded that our server was mostly used by players on games which promoted small groups. However, because determining the game being played is impossible from our logs, and because the server was advertised to a wide variety of sources, we believe our results are general enough to at least apply to MVC for games in general.

4.4. SESSION CHARACTERISTICS

We define a session as the period from when a user logs into the server until they log out of the server. These can be determined finding the *login* and *logout* entries in the TeamSpeak log file. Note that entries where the user was seen logging in more than once without logging out were not considered in our measurements.

We recorded 7,749 sessions, including the packets that were sent to and from the server and how long users were logged into TeamSpeak. *On average, we observed 86.1 logins per day from 826 individual users.* To understand this data further, we calculated the session times and generated a CDF as shown in Figure 5.

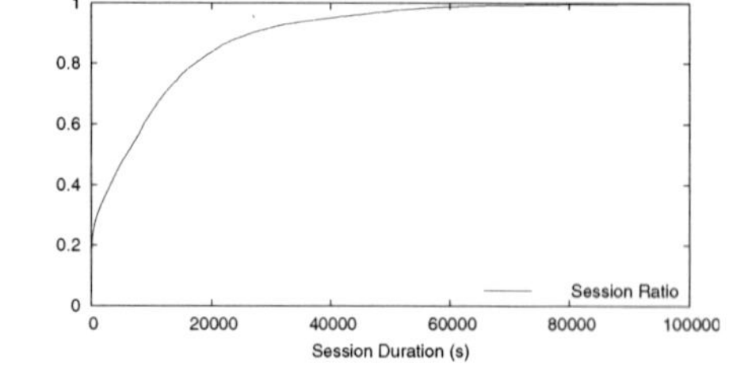

Figure 5: *Sessions Length CDF:* We see that of the 7,749 sessions we recorded, half of these sessions were less than 5000 seconds (1.3 hours). A small fraction of these (a few hundred) were over 30,000 seconds (8 hours).

Our calculations show that the shortest sessions were less than one second while the longest session was over 69 hours! However, as Figure 5 shows, for 20% of the sessions, users stayed less than 1/2 hour. In addition, for 20% of the sessions, users stayed for more than 5 hours. Thus, 60% of the sessions fell somewhere between 1/2 hour and 5 hours.

For the small fraction of sessions that were greater than 8 hours, we hypothesize that users simply did not log out of the TeamSpeak server when they were done. In the future, we would like to look at the correlation of session time with input traffic to see if long lived or very short sessions are actually sending and receiving voice traffic.

The characteristic of our curve is similar to both that can be found in [7] and in [16]. However, both of these papers analyze network traffic in on-line games, one of them focuses on a First Person Shooter (FPS) game whereas the other one focuses on a Massively Multiplayer Online Game (MMOG). On the one hand this fact validates our results but on the other it shows that it is nearly impossible to conclude what type of game is played by the users analyzing only the characteristics of the data and not the content of it.

Table 2: *Cleaning the data sets:* The effect of removing extreme outliers with the cleaning procedure on inter-talkspurt, talkspurt, and silence periods data sets.

	Inter-talkspurt arrival	**Talkspurt**	**Silence**
Original data size	188,225 (100%)	188,313 (100%)	186,626 (100%)
Filtered data size	187,884 (99.82%)	188,158 (99.92%)	186,626 (100%)
Deleted data size	341 (0.18%)	155 (0.08%)	0 (0.00%)

In the next measurements, we matched IP addresses with sessions to determine how many unique IP addresses logged into the system. In essence, we would like to determine how frequently a user logs into and uses the TeamSpeak server. We calculated the CDF of the ratio of logins versus the number of logins as illustrated in Figure 6.

Our results indicate that 40% of the users logged into the TeamSpeak server only once, while only 17% logged into it regularly (i. e. at least once every week on average). However, this result might be biased due to the fact that some users may be using DHCP to receive their IP addresses when they use the Internet. Thus, multiple IP addresses may refer to the same user and the total number of users we saw may be fewer. Although broadband Internet users usually keep their IP addresses for several days or even weeks, we are currently investigating ways to determine the correct identity of users.

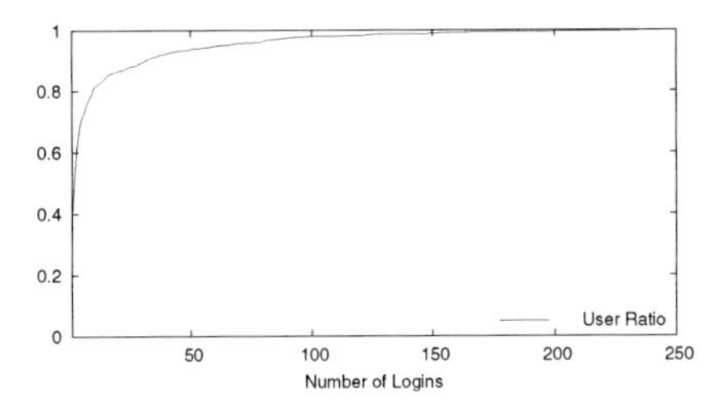

Figure 6: *Login Count CDF:* 40 % of all the IP addresses that logged to our server were unique. This is probably due to the fact that DHCP was used to assign their addresses. 17% appeared to log into the server at least once a week on average.

4.5. MEASURED VOICE PATTERNS

Voice patterns in multiparty voice communication consist of talkspurts (on periods) and silence (off periods). We measured these and the inter-talkspurt arrival time to characterize voice patterns. TeamSpeak uses 100ms long frames, therefore the shortest talkspurt in our case is 100ms. To be consistent, the smallest measureable silence period must also be 100ms. The inter-talkspurt arrival time is measured as the time between any two in-sequence talkspurts observed by the server. Since the smallest talkspurt is 100ms and the smallest silence period is 100ms, then any inter-talkspurt arrival time that is at least 200ms for a given user is interpreted as silence. Note that, if we have multiple users using the server at the same time the talkspurts can overlap and thus the inter-talkspurt arrival time can be shorter than the talkspurt itself.

In order to measure the voice patterns, we captured the voice packets during the peak periods (7pm–9pm). After sorting and analyzing the data we realized that our data points did not fit on a linear curve. Therefore, we identified the extreme outliers using the method described in Section 3.2 and removed them from our data set. When we applied the cleaning procedure to the inter-talkspurt arrival times, we removed 341 data points. Table 2 shows the results of cleaning the inter-talkspurt arrival times.

Figure 7 plots the inter-talkspurt arrival times seen at the server during the peak periods. The majority (90%) of the inter-arrival times is less than 7.65 sec. However, the remaining 10% of the data forms a tail which stretches to 536.83 seconds. Note that we only include the first 100 seconds and use a log-log scale in the graph so that the CDF can be seen more clearly.

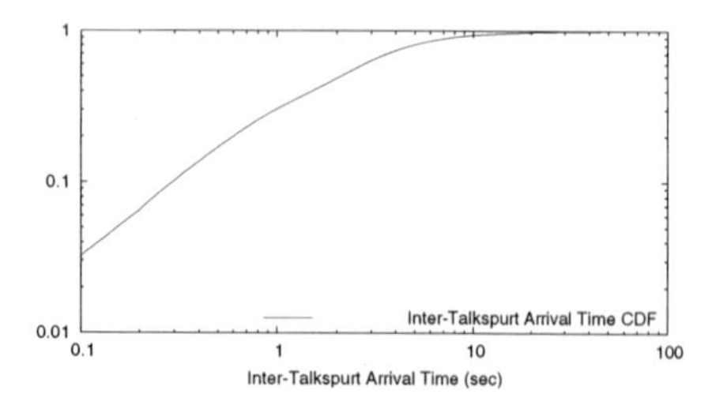

Figure 7: *Inter-talkspurt arrival time:* The majority (90%) of the inter-talkspurt arrival times are less than 7.65 sec. The resulting CDF appears to be exponential in nature.

We collected the talkspurts and silence periods for each of the users during the peak period. We then merged these sets into a single data set and found that the data was non-linear. We transformed it and deleted the extreme outliers with the results listed in Table 2.

In Figures 8 and 9, we plot the CDFs of the talkspurts and silence periods, respectively. Both CDFs appear to follow an exponential distribution, which we explore further in Section 5. The expected value of the talkspurts is much lower than the expected value of the silence periods. 90% of the talkspurts are shorter than 5.40 sec, whereas the same measure for the silence periods is 70.11 sec, which is around an order of magnitude higher. This implies that the users tend to listen more than they talk. After the filtering process, our lowest talkspurt value was .1 sec and our highest value was 96.46 sec. This can be seen in Figure 8.

When we analyzed the silence periods, the filtering process did not affect our data set (see Table 2). This is due to the fact that the expected value of our exponential-like curve was higher and thus the IQR was broader. In addition, because our measurements were only performed during the peak period, the silence periods have an upper bound of 3 hours (or 10,800 secs). The silence period data set ranged from .1 sec (the minimum possible silence period) to 7036.95 sec (almost 2 hours). However, in order to examine the curve of the CDF better, we only include the first 1000 seconds in Figure 9. In the next section, we model the data sets mathematically and discover that both the talkspurt and silence periods are better modeled by Weibull distributions.

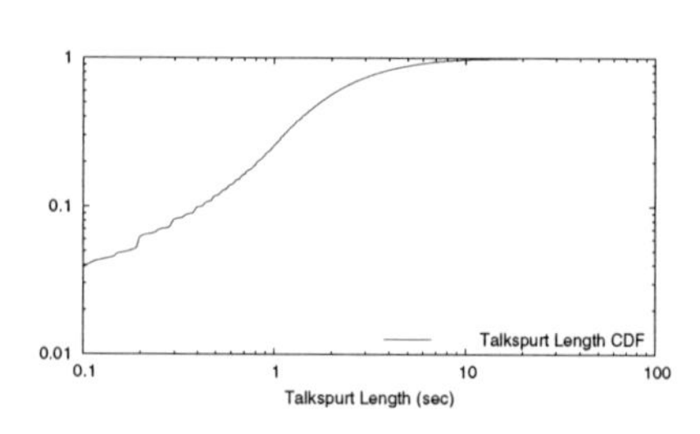

Figure 8: *Talkspurts:* The majority (90%) of the talkspurts are less than 5.40 sec.

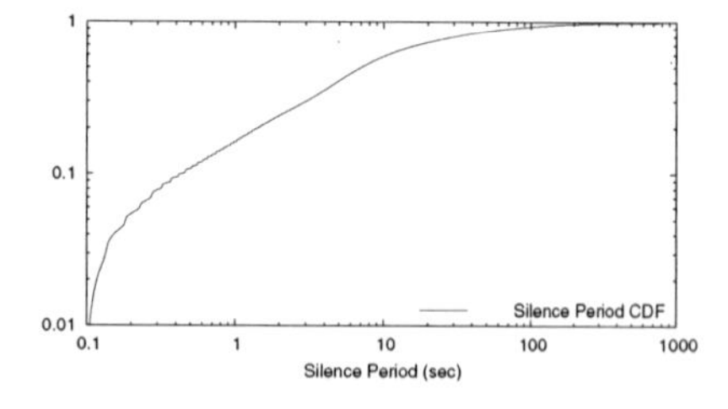

Figure 9: *Silence periods:* The majority (90%) of the silence periods are less than 70.11 sec, which is much higher than the talkspurts.

5. MODELING MULTIPARTY VOICE COMMUNICATION

We now model multiparty voice communication. We have three primary factors that we need to model mathematically: talkspurts, silence and group sizes. With these models, we can simulate and predict the characteristics of multiparty voice communication, regardless of whether a client/server, peer-to-peer or hybrid architecture is used.

5.1. METHODOLOGY

Initially, we thought that the data appeared to follow some kind of exponential distribution, but as we analyzed the data further, we discovered that it fits a Weibull distribution better. Note that this differs from previous research in classical telephony and VoIP conversations which showed that the data followed an exponential distribution.

In order to model the conversations, we first estimated the parameters of the exponential and Weibull distributions. We looked at other distributions, but found that these two distributions had the best fit with our data. We then validated our estimation by calculating the mean and standard deviation of the residuals and by using the λ^2 test.

5.2. PARAMETER ESTIMATION AND ERROR CALCULATION

For parameter estimation, we used the least-squares method to minimize the square of the sum of the residuals for both the Exponential and Weibull distributions.

In order to justify the correctness of our estimation, one could perform a goodness-of-fit test. However, traditional tests, such as Chi-square (χ^2) and Kolmogorov-Smirnov (KS) are not suitable for data from Internet traffic [12]. Moreover, these tests are biased against large data sets [6], such as the ones that we have.

We use two methods to determine if the data fits a particular distribution. After we have used the least squares method to estimate the parameters for a distribution, we plot the residuals and examine their mean and standard deviation. These values give us an idea of how well our model predicts the data. In addition, we use the λ^2 method as a discrepancy tool [13]. We describe how we used the λ^2 method and how we binned our data in Subsection 5.3. With the λ^2 method, we can compare the fit between two possible distributions.

5.3. USING λ^2 FOR NETWORK MODEL EVALUATION

The quantity λ^2 is the discrepancy between an actual and an assumed statistical model, which is the measure of the goodness-of-fit of the estimated curve. However, this method can be applied to data in different ways. Here, we present the details of how we applied it.

The λ^2 metric is defined as follows:

$$\lambda^2 = \frac{\chi^2 \quad K \quad df}{n - 1} \tag{2}$$

where n is the total number of datapoints and df is the number of degrees of freedom of the test.

$$\chi^2 = \sum_i \frac{(O_i - E_i)^2}{E_i} \tag{3}$$

and

$$K = \sum_i \frac{|O_i - E_i|}{E_i} \tag{4}$$

Since this discrepancy is based on Pearson's χ^2 test, it requires the binning of the data. Here O_i is the observed number of datapoints in bin i and E_i is the estimated number of datapoints in bin i. Please note that not just χ^2 but K is also dependent on the number of bins and therefore determining this parameter can be crucial. If the parameter is too large, the estimate will be too rough; on the other hand if the parameter is too small than the distribution of the datapoints will be too smooth, equivalent statistically to imprecise estimation.

In our paper we used $1 + 2 \times 2 \times \log_{10} n$ equiprobable bins [9]. If this was not a whole number we took the floor of it. This method ensures that if we have at least one datapoint the denominator of neither χ^2 nor K can be zero. Our experience is that these parameters were accurate and worked well with our data because they were in accordance with what we expected based on the graphs.

5.4. MODELING TALKSPURTS AND SILENCE

To model the talkspurts and silence periods, we looked at the packets sent and received during the peak period on the server (from 7pm to 9pm). We focus on this period because the model needs to be able to predict the behavior under peak loads. After looking at the data, graphed in Figure 8, we hypothesized that the data followed the exponential distribution.

Table 3: *Experimental values:* The Mean, Min and Max are calculated from the data sets. Using our parameter estimation methods, we calculated the parameters for the CDFs of the exponential and Weibull distributions. The λ^2 values are the results of using the λ^2 test to determine the accuracy of our fit (smaller is better). For both the talkspurt and silence data sets, the Weibull distribution is a better fit.

	Talkspurt	**Silence**
Mean	2.74s	35.90s
Min	0.1s	0.1s
Max	96.46s	7036.95s
Exponential estimated parameters	$\lambda = 0.4185$	$\lambda = 0.0877$
Weibull estimated parameters	$\lambda = 2.3002$ $k = 1.1846$	$\lambda = 13.5275$ $k = 0.6168$
λ^2−test for exponential	0.0999	0.2739
λ^2−test for Weibull	0.0769	0.0636

Table 4: *Residuals from Model:* The max, min, and standard deviation of the residuals between the modeled CDFs and the talkspurts and silence periods.

	Talkspurt	**Silence**
Max	0.0401	0.0269
Min	-0.0350	-0.0382
Std.Dev.	0.0190	0.0180

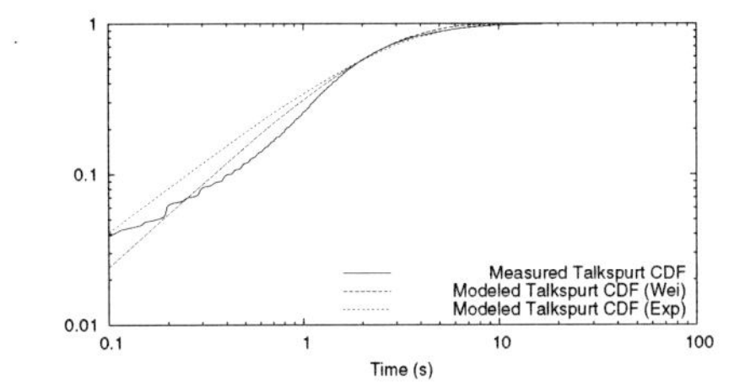

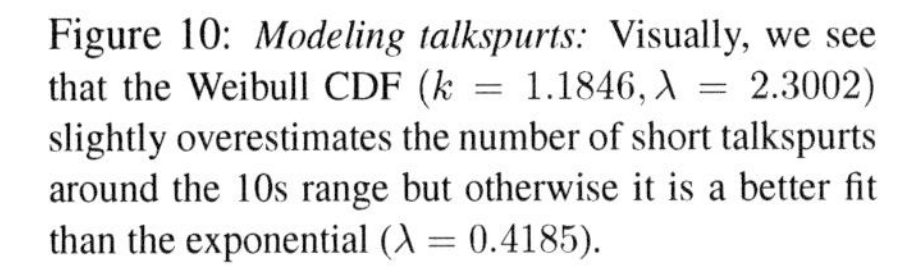
Figure 10: *Modeling talkspurts:* Visually, we see that the Weibull CDF ($k = 1.1846, \lambda = 2.3002$) slightly overestimates the number of short talkspurts around the 10s range but otherwise it is a better fit than the exponential ($\lambda = 0.4185$).

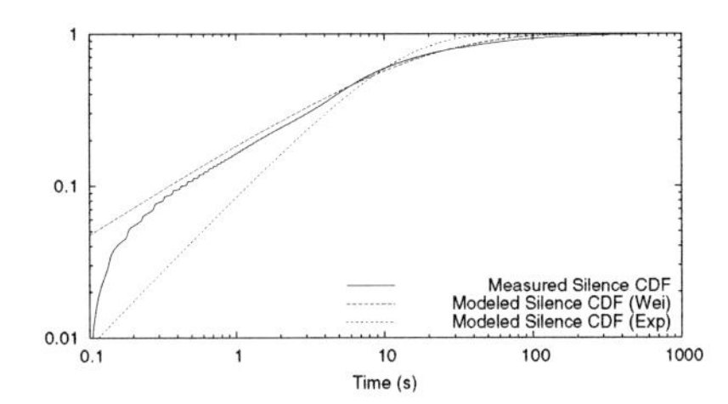

Figure 11: *Modeling silence:* We model the silence period using a Weibull CDF ($k = 0.6168, \lambda = 13.5275$). We see that like the talkspurts, the Weibull distribution fits better than the exponential distribution ($\lambda = 0.0877$).

Our first attempt at modeling the talkspurt and silence periods was to try an exponential distribution. Table 3 lists the means and parameters we estimated using the least-squares method for the exponential distribution.

We then tried the Weibull CDF which is a generalization of the exponential distribution. We estimated the parameters for the Weibull distribution for both the talkspurts and silence periods (Table 3). We then plotted the talkspurt and silence data sets along with both exponential and Weibull CDFs using their estimated parameters. The talkspurt graph with its models can be seen in Figure 10. Visually, the Weibull CDF appears to be a better fit for the talkspurt data set. We plotted the residuals (not shown) to examine their mean and standard deviations and summarize them in Table 4.

We then used the λ^2 test on the CDFs and validated that the Weibull CDF fits better than the exponential as shown in Table 3. Thus, *unlike prior results which showed that an exponential distribution better modeled talkspurts, we found that the Weibull CDF more accurately models the talkspurts of multiparty voice communication.*

For the silence periods, we repeated our method of plotting the data set with both the exponential and Weibull CDFs and their estimated parameters, as shown in Figure 11. To further validate the results, we plotted the residuals (not shown), which are the differences between the predicted values and observed values. The residuals shows us that the model is off by at most 4%, with a standard deviation less than .02 as shown in Table 4. Using the λ^2 test, we see that our estimated Weibull CDF is indeed a better fit than the exponential distribution (Table 3). Thus, *the silence periods are more accurately modeled with Weibull CDF for multiparty voice communications.*

5.5. GROUP EFFECTS ON TALKSPURT AND SILENCE

Besides the talkspurt and silence distributions, we wanted to understand how group sizes affect these distributions. We hypothesized that as the number of people in a group increased, the mean talking time decreased while the mean silence time increased. To study this, we plotted the mean talkspurt and silence times versus the group sizes observed during our measurement period.

As TeamSpeak does not use a group identifier in the messages, it is impossible to identify the groups with 100% accuracy. However, for modeling the behavior of the groups with different sizes it is not essential to associate the messages to a particular group. Simply knowing the size of the group that a message was sent to would be sufficient if this method was also capable of grouping the silent periods based on the group size. Thus, we counted the number of replications for each of the incoming messages from a given user. Next, we used this group size to determine the group size for the following silence period. This way we could associate a group size to both the talkspurts and the silence periods. The only time when our method fails is when a player leaves or joins a group during a silence period. However, this event is very unlikely and therefore our solution is capable of providing an accurate result.

Figure 12 shows the mean talkspurt and silence times versus the group size. We only show groups of up to 8 people due to the fact that while we did observe groups with up to 24 people, the number of data points in these larger groups were too few to be statistically meaningful.

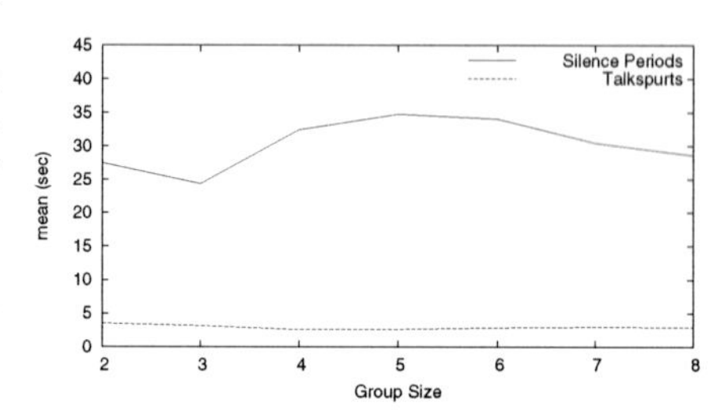

Figure 12: *Talkspurts and silence periods among groups:* Note that the mean talkspurt and silence times are fairly constant.

Looking at this graph, we see that the mean talking and mean silence time do not change significantly, regardless of the group size, contradicting our hypothesis. To investigate this unexpected result, we ran a script which looked at the number of people talking in a group and found that as the group size increases, the number of completely silent people increases (e.g., they only have headphones, but not a mic to speak on). In essence, *a single group appears to support a maximum amount of conversation, regardless of the group size.* We expect that future architectures and codecs may be able to take advantage of this information.

6. CONCLUSION AND FUTURE WORK

We have presented the first work that examines the characteristics of multiparty communication for games. While VoIP has been successful for point-to-point communication, and research has looked at the feasibility of VoIP over the Internet, our work is the first to address multiparty voice communications on the Internet.

Our results show familiar and new trends. First, as we modeled the talkspurts and silence periods, we found that both types of data fit a Weibull CDF, which differs from previous work on traditional telephony and VoIP that shows talkspurts following exponential distribution. Moreover, we showed that the length of the talkspurts and silence are always the same regardless either of the game played or the group size. On the other hand, the distribution of our daily traffic was similar to other works in both games and VoIP, where server usage peaked during evening hours and on weekends.

Finally, human protocols seem to be at work here as our measurements indicate. The increase in group sizes does not increase the amount of input traffic linearly, though output traffic is necessarily linear in the number of packets received. This is simply due to the fact that humans best process voice information when only one person is talking at the same time. Thus, if more than one person starts talking, other speakers naturally back-off and wait for their turn.

As for future work, we plan on using our models for simulation of client/server and peer-to-peer multiparty voice communication systems.

REFERENCES

[1] BASET, S., AND SCHULZRINNE, H. An analysis of the Skype peer-to-peer internet telephony protocol. In *Proceedings of IEEE Infocom* (April 2006), pp. 1–11.

[2] BORELLA, M. Source models of network game traffic. *Computer Communications 23*, 4 (February 2000), 403–410.

[3] BRADY, P. A statistical analysis of on-off patterns in 16 conversations. *Bell Systems Technical Journal 47*, 1 (January 1968), 73–91.

[4] CHEN, K.-T., HUANG, C.-Y., HUANG, P., AND LEI, C.-L. Quantifying Skype user satisfaction. In *Proceedings of ACM SIGCOMM* (2006), pp. 399–410.

[5] FÄRBER, J. Network game traffic modelling. In *Proceedings of the 1st workshop on Network and system support for games* (2002), pp. 53–57.

[6] GLESER, L. J., AND MOORE, D. S. The effect of dependence on chi-squared and empiric distribution tests of fit. *Annals of Statistics 11*, 4 (1983), 1100–1108.

[7] HENDERSON, T., AND BHATTI, S. Modelling user behaviour in networked games. In *MULTIMEDIA '01: Proceedings of the Ninth ACM International Conference on Multimedia* (New York, NY, USA, 2001), ACM Press, pp. 212–220.

[8] JIAN, W., AND SCHULZRINNE, H. Analysis of on-off patterns in VoIP and their effect on voice traffic aggregation. In *Proceedings of Computer Communications and Networks* (Oct. 2000), pp. 82–87.

[9] LARSON, H. J. *Statistics: An introduction.* Wiley, New York, NY, USA, 1975.

[10] MARKOPOULOU, A. P., TOBAGI, F. A., AND KARAM, M. J. Assessing the quality of voice communications over internet backbones. *IEEE/ACM Trans. Netw. 11*, 5 (2003), 747–760.

[11] PAPP, G., AND GAUTHIERDICKEY, C. Characterizing multiparty voice communication for multiplayer games (extended abstract). In *to appear in ACM Sigmetrics* (June 2008).

[12] PAXSON, V. End-to-end routing behavior in the internet. *IEEE/ACM Trans. Netw. 5*, 5 (1997), 601–615.

[13] PEDERSON, S. P., AND JOHNSON, M. E. Estimating model discrepancy. *Technometrics 32*, 3 (1990), 305–314.

[14] PITTMAN, D., AND GAUTHIERDICKEY, C. A measurement study of virtual populations in massively multiplayer online games. In *Proceedings of ACM NetGames* (September 2007).

[15] SRIRAM, K., AND WHITT, W. Characterizing superposition arrival processes in packet multiplexers for voice and data. *IEEE Selected Areas in Communications 4*, 6 (1986), 833–846.

[16] SVOBODA, P., KARNER, W., AND RUPP, M. Traffic analysis and modeling for world of warcraft. *2007. ICC '07. IEEE International Conference on Communications* (24-28 June 2007), 1612–1617.

[17] WU-CHANG FENG, CHANG, F., WU-CHI FENG, AND WALPOLE, J. Provisioning on-line games: A traffic analysis of a busy counter-strike server. In *Internet Measurement Workshop* (2002).

Gerhard MÜNZ*, Georg CARLE*

APPLICATION OF FORECASTING TECHNIQUES AND CONTROL CHARTS FOR TRAFFIC ANOMALY DETECTION

In this paper, we evaluate the capability to detect traffic anomalies with Shewhart, CUSUM, and EWMA control charts. In order to cope with seasonal variation and serial correlation, control charts are not applied to traffic measurement time-series directly, but to the prediction errors of exponential smoothing and Holt-Winters forecasting. The evaluation relies on flow data collected in an ISP backbone network and shows that good detection results can be achieved with an appropriate choice and parametrization of the forecasting method and the control chart. On the other hand, the relevance of the detected anomalies for the network operator mainly depends on the monitored metrics and the selected parts of traffic.

1. INTRODUCTION

In control engineering, monitoring mechanisms are deployed to observe the properties or behavior of a system and raise an alarm if an important parameter runs out of the range of sound operation. One monitoring goal is the detection of unexpected changes in characteristic properties of the system because such changes may be indications of failures, malfunctions and wearout. The detection must be fast to enable manual intervention, recalibration, or exchange of erroneous system components before severe consequences happen. Network monitoring pursues similar objectives. One aspect is the identification of anomalous traffic behavior which can be a sign of network failures or abuses, for example due to worm or attack traffic.

Change detection methods consider time-series of measurement values and search for points in time at which statistical properties of the measurements change abruptly, i.e. "instantaneously or at least very fast with respect to the sampling period of the measurements" [2]. Before and after the change, the monitored statistical properties are assumed to show no or only little variation. Under these conditions, even small changes can be detected with high probability if they persist for a long duration.

In general, the more a priori knowledge is available, the easier it is to detect changes with high accuracy. For example, parametric methods have more power than non-parametric methods, which means that they allow us to detect more true anomalies at the same false alarm level (i.e., probability of an alarm in absence of any significant change). However, if the model assumption is incorrect,

*Network Architectures and Services – Institute for Informatics, Technische Universität München, Germany,
e-mail: {muenz|carle}@net.in.tum.de

parametric methods lose their decisive power and may lead to wrong decisions. In the case of traffic anomaly detection, we cannot assume that the monitored variables follow a specific distribution, thus the detection should be non-parametric, or at least robust against non-normality. Moreover, changes should be detected very quickly (i.e., online) without requiring any a priori knowledge about their magnitude since such information is usually not available, either.

A practical solution to statistical online change detection are control charts [16]. In a control chart, mean and variability of a monitored variable are characterized by a centerline (CL), an upper control limit (UCL), and a lower control limit (LCL). A change is detected if the measured value exceeds one of the control limits. This decision can be formalized as a hypothesis test with null hypothesis H_0 assuming no significant change and alternative hypothesis H_1 suggesting the opposite.

In this paper, we evaluate the applicability of control charts to the problem of traffic anomaly detection. More precisely, we analyze time-series of byte, packet, and flow counts which can be easily obtained from routers via SNMP (Simple Network Management Protocol), IPFIX [8], or Cisco NetFlow [9]. These measurement time-series are subject to systematic changes, in particular seasonal variation. In addition, dependencies may exist between subsequent observations, noticeable as serial correlation. Both, systematic changes as well as serial correlation, need to be accounted for because most control charts are designed for independent and identically distributed observations. Useful tools are forecasting techniques which predict future values based on what has been observed in the past. In the optimal case, the prediction errors are small, random, and uncorrelated as long as the behavior of the monitored variable does not change. Hence, we can identify changes in the original variable by applying control charts to the prediction errors.

The main contribution of this work consists of a comparison of three different control charts and two different forecasting techniques. The considered control charts are the Shewhart control chart, the CUSUM (cumulative sum) control chart, and the EWMA (exponentially weighted moving average) control chart, which are commonly used in process monitoring. With respect to forecasting, we choose exponential smoothing and Holt-Winters forecasting, two self-adaptive and robust forecasting techniques which are suitable for our purposes. In contrast to many existing publications, our evaluation is not based on synthetically generated anomalies. Instead, we apply our methods to real flow data collected in the backbone network of an Internet Service Provider (ISP) and assess the relevance of the detected anomalies by examining their causes. The evaluation shows that a combination of Shewhart control chart and exponential smoothing enables good detection results under various conditions.

Sections 2 and 3 provide the theoretical background of the deployed control charts and forecasting techniques. Subsequently, in Section 4, we evaluate the capability to detect traffic anomalies with the described methods. Section 5 surveys related approaches of using control charts for traffic anomaly detection before Section 6 concludes this paper.

2. CONTROL CHARTS

A typical control charts contains a center line (CL) representing the average value of the monitored random variable Y under normal conditions. Above and below the center line, the upper and lower control limit (UCL, LCL) define the range of normal variation or in-control state. The decision function detects a change (or out-of-control state) if the measured value y lies outside this range.

The statistical properties of control charts can be deduced from the theory of sequential probability ratio tests (SPRT). If the distribution before and after the change are known, the decision function can be converted into a condition for the log-likelihood ratio of the observation y:

$$s(y) = \log \frac{p_{\Theta_1}(y)}{p_{\Theta_0}(y)}$$

where $p_\Theta(y)$ is the probability density function of Y with parameter Θ. Θ_0 and Θ_1 are the parameter values before and after the change. If $s(y)$ is positive, the monitored random variable more likely conforms to the distribution after change than to the distribution before change. Hence, we can define a threshold h for $s(y)$ to reject the null hypothesis $H_0 : \Theta = \Theta_0$ and accept the alternative hypothesis $H_1 : \Theta = \Theta_1$ at a given level of significance. The level of significance corresponds to the probability of a false alarm.

If Y is normally distributed with constant variance σ^2 and means μ_0 and μ_1 before and after change, $s(y)$ becomes:

$$s(y) = \frac{\mu_1 - \mu_0}{\sigma^2} \left(y - \frac{\mu_1 + \mu_0}{2} \right)$$

If $\mu_1 > \mu_0$, $s(y) > h$ is equivalent to the decision function:

$$y > \mu_0 + L\sigma \quad \text{with } L = \frac{h\sigma}{\mu_1 - \mu_0} + \frac{\mu_1 - \mu_0}{2\sigma}$$

In this equation, the correspondence to the control chart is obvious: μ_0 is the center line and $\mu_0 + L\sigma$ the upper control limit.

As the variance of a single observation is quite high, change detection methods often consider a sequence of observations $\{y_t | t = a, \ldots, b\}$ to increase the power of the hypothesis test. Under the condition that the observations are independent, the log-likelihood ratio of the sequence is:

$$s(y_a, \ldots, y_b) = \log \frac{\prod_{t=a}^{b} p_{\Theta_1}(y_t)}{\prod_{t=a}^{b} p_{\Theta_0}(y_t)} = \sum_{t=a}^{b} s(y_t)$$

A hypothesis test based on $s(y_a, \ldots, y_b)$ corresponds to a control chart for a test statistic that is calculated from $y_a, \ldots y_b$. An example is the average value $\bar{y}$, which has an important property: regardless of the distribution of Y, $\bar{y}$ is approximately normally distributed if calculated from a large number of observations, thanks to the central limit theorem.

In the following subsections, we introduce three different control charts, namely the Shewhart control chart, the CUSUM control chart, and the EWMA control chart. For more detailed information, we refer to the text books of Montgomery [16] and Basseville [2].

2.1. SHEWHART CONTROL CHART

The Shewhart control chart [26] defines UCL, CL, and LCL for a statistic calculated from N observations $y_{(l-1)N+1}, \ldots, y_{lN}$. An example statistic is the average value $\bar{y}_l$, which is appropriate for detecting changes in the mean:

$$\bar{y}_l = \frac{1}{N} \sum_{t=(l-1)N+1}^{lN} y_t \quad \text{where} \quad l = 1, 2, \ldots$$

If the observations are independent and identically distributed with mean μ_0 and variance σ^2, $\bar{y}_l$ is an unbiased estimator of μ_0 with variance σ^2/N. Hence, the upper and lower control limits can be defined in the form $\mu_0 \pm L\sigma/\sqrt{N}$ with tuning parameter L. An alarm is raised if $\bar{y}_l$ passes one of the control limits. As already mentioned, $\bar{y}_l$ is approximately normally distributed for large N, thus the control limits for a given false alarm probability α are $\mu_0 \pm \Phi(1-\alpha/2)\sigma/\sqrt{N}$. However, this approximation does not hold for small N or if the observations are serially correlated.

A special case is $N = 1$, the so-called Shewhart control chart of individuals. This chart compares individual observations against the control limits. Obviously, the central limit theorem does not apply, thus the distribution of Y needs to be known exactly in order to define precise limits for a given false alarm probability.

2.2. CUSUM CONTROL CHART

The CUSUM control chart (also called CUSUM algorithm) [18] is based on the fact that $S_t = s(y_1, \ldots, y_t)$ has a negative drift under normal conditions and a positive drift after a change. The CUSUM decision function g_t compares the increase of S_t with respect to its minimum to a threshold h:

$$g_t = S_t - \min_{1 \le i \le t} S_i = \max\left(0, s(y_t) + g_{t-1}\right) = [g_{t-1} + s(y_t)]^+ \ge h \quad ; \; g_0 = 0$$

An alarm is raised if g_t exceeds the threshold h. To restart the algorithm, g_t must be reset to zero.

From the view of hypothesis testing, the CUSUM control chart repeatedly performs an SPRT where each decision considers as many consecutive observations as needed to accept either H_0 or H_1. The CUSUM control chart implicitly starts a new run of SPRT if H_0 has been accepted, and stops with an alarm in the case of H_1. The threshold h allows trading off the mean detection delay and the mean time between false alarms. If the distribution of Y is unknown, the log-likelihood ratio $s(y_t)$ must be replaced by a statistic $u(y_t)$ with comparable properties: the expectation value of $u(y)$ must be negative under H_0 and positive under H_1 This variant is often called non-parametric CUSUM algorithm.

An appropriate statistic for detecting positive shifts in the mean is $u^+(y) = y - (\mu_0 + K)$. K is called reference value. In order to detect negative shifts as well, we need a second statistic $u^-(y) = (\mu_0 - K) - y$. As a result, we get two decision functions:

$$g_t^+ = [g_{t-1} + y_t - (\mu_0 + K)]^+ \ge h \quad ; \quad g_t^- = [g_{t-1} + (\mu_0 - K) - y_t]^+ \ge h$$

Typical settings are $K = \sigma/2$ and $h = 4\sigma$ or $h = 5\sigma$, where σ is the standard deviation of Y_t [16].

Compared to the Shewhart control chart, CUSUM detects small but persistent changes with higher probability because little effects accumulate over time. Brodsky and Darkhovsky [5] studied the properties of the non-parametric CUSUM algorithm for a specific family of exponential distributions of $u(y)$. For this distribution family, the detection delay reaches the theoretic minimum if the mean time between false alarms goes to infinity. As we will discuss in Section 5, several existing publications refer to this proof of optimality although the specific requirements are not fulfilled in general.

2.3. EWMA CONTROL CHART

The EWMA control chart (c.f. [23,24]) relies on exponential smoothing of observations. Given the smoothing constant λ $(0 < \lambda < 1)$,

$$z_t = \lambda y_t + (1-\lambda) z_{t-1} = \lambda \sum_{i=0}^{t-1} (1-\lambda)^i y_{t-i} + (1-\lambda)^t z_0$$

is a weighted average of all observations up to time t. The initial value is the expected mean under H_0: $z_0 = \mu_0$. If the observations are independent and identically distributed with variance σ^2, the variance of z_t approaches $\frac{\lambda}{2-\lambda}\sigma^2$ for $t \rightarrow \infty$, which allows the definition of control limits for z_t:

$$\mu_0 \pm L\sigma\sqrt{\frac{\lambda}{2-\lambda}}$$

λ and L are design parameters of the EWMA control chart. Popular choices are $2.6 \leq L \leq 3$ and $0.05 < \lambda < 0.25$, where smaller λ allow detecting smaller shifts [16].

The EWMA control chart has some interesting properties [16]. It can be tuned to achieve approximately equivalent results as the CUSUM control chart. Secondly, it is quite robust against non-normal distributions of Y, especially for small values of λ (e.g., $\lambda = 0.05$). Finally, after adjusting the control limits, the EWMA control chart still performs well in the presence of low to moderate levels of serial correlation in Y_t.

3. RESIDUAL GENERATION BY FORECASTING

Common to all the control charts presented in Section 2 is the assumption that the observations are independent and identically distributed under normal conditions. This corresponds to the output of a stationary random process that generates uncorrelated values. Such a process is also called pure random process.

There are various reasons why traffic measurement data exposes significant deviation from the output of a pure random process. Non-stationarities result from trends as well as dependencies on the time of day, the day of the week etc. Serial correlation is caused by internal network states which cannot change arbitrarily from one instant in time to the next. For example, the number of packets in the network evolves according to a birth-death process depending on the arrival times and processing times.

We can identify systematic changes in the mean or variance by visually inspecting the measured values over time. Systematic changes as well as serial correlation also have an impact on the sample autocorrelation, which is calculated as follows:

$$r_\tau = \frac{\sum_{i=1}^{N-\tau}(y_t - \bar{y}_t)(y_{t+\tau} - \bar{y}_t)}{\sum_{i=1}^{N}(y_t - \bar{y}_t)^2}$$

In the above equation, N is the number of observations and τ the lag between two instances of time. If r_τ is not decreasing with increasing τ, or if it shows periodic oscillation, the observations do not resemble the output of a stationary random process. In the case of a pure random process, the 95% confidence interval of r_τ is $[-1/N - 2/\sqrt{N}; -1/N + 2/\sqrt{N}]$ for all τ. Hence, if a non-negligible number of r_τ's lie outside this range, the process is not purely random.

Time-series analysis allows modeling and removing systematic changes and serial correlation with help of the Box-Jenkins approach [4] . However, fitting an accurate ARIMA (autoregressive integrated moving average) model is difficult and requires a long series of anomaly-free observations. Therefore, robust forecasting methods based on exponential smoothing are preferred [7], especially for online applications. Forecasting relies on the assumption that the temporal behavior observed in past observations persists in the near future. Hence, an unusually large prediction error is an indicator of a change in the monitored random variable. The prediction errors are also called residuals because they represent the variability not explained by the forecasting model.

In the following subsections, we present two popular forecasting techniques that we will use in Section 4 for residual generation: exponential smoothing and Holt-Winters forecasting. In order to define appropriate limits for the control charts, we need to estimate the standard deviation of the residuals under normal conditions. How this can be achieved is explained in Section 3.3.

3.1. EXPONENTIAL SMOOTHING

Exponential smoothing allows predicting future values by a weighted sum of past observations:

$$\hat{y}_{t+1} = \alpha \sum_{i=2}^{t} (1-\alpha)^i y_i + (1-\alpha)^t y_1 = \alpha y_t + (1-\alpha)\hat{y}_t$$

This is the same exponentially weighted moving average as used in the EWMA control chart. The distribution of the weights is geometric and gives more weight to recent observations. Forecasting according to the above equation is optimal for an infinite-order MA (moving average) process, which is equivalent to an ARIMA(0,1,1) process [7]. Yet, exponential smoothing is very robust and also provides good forecasts for other trendless and non-seasonal time-series. The optimal value for α can be approximated by trying different values and choosing the one with the smallest residual sum of squares.

3.2. HOLT-WINTERS FORECASTING

Holt-Winters forecasting combines a baseline component L_i with a trend component T_i and a seasonal component I_t:

$$\hat{y}_{t+1} = L_t + T_t + I_t$$

L_t, T_t, and I_t are recursively updated according to the following equations:

$$L_t = \alpha(y_t - I_{t-s}) + (1-\alpha)(L_{t-1} + T_{t-1})$$

$$T_t = \beta(L_t - L_{t-1}) + (1-\beta)T_{t-1}$$

$$I_t = \gamma(y_t - L_t) + (1-\gamma)I_{t-s}$$

α, β, and γ are smoothing parameters which have to be set to appropriate values in the range $(0,1)$. s is the length of one season counted in time intervals. The above equations include an additive seasonal component. Alternatively, the seasonal component can also be modeled in a multiplicative way. For more details, we refer to [7] and the references therein.

3.3. CONTROL LIMITS AND STANDARD DEVIATION ESTIMATORS

As we have seen in Section 2, control limits are usually defined relatively to the standard deviation σ of the monitored random variable. If we apply control charts to residual time-series of prediction errors $\epsilon_t = y_t - \hat{y}_t$, the standard deviation has to be estimated. We could calculate the sample variance from a finite set of past residuals. However, this estimation is very sensitive to outliers and does not reflect dynamic changes in the variance. Therefore, we make use of a moving estimator which is based on exponential smoothing. For a given mean μ, the exponentially weighted mean square error (EWMS) is a variance estimator:

$$\hat{\sigma}_t^2 = \rho(\epsilon_t - \mu)^2 + (1-\rho)\hat{\sigma}_{t-1}^2$$

Since the mean of the residuals is approximately zero under normal conditions, we can set $\mu = 0$ in the above equation.

4. EVALUATION

We evaluated the capability to detect traffic anomalies with help of the forecasting techniques and the control charts presented in the previous sections. Our evaluation is based on traffic measurement data collected in the Gigabit backbone network of a regional ISP between September 7 and October 25, 2006. The operation area of the ISP covers parts of Saarland, Rhineland-Palatinate, Hesse (all federal states in Germany), Luxembourg, and Belgium. At measurement time, the offered services ranged from server hosting and colocation to VPNs and modem, ISDN, and DSL dial-in service. Customers were corporate clients, local carriers, roaming providers, and small and medium enterprises. The measurements were performed at a router using unsampled Cisco Netflow.v5 with active and idle flow timeouts set to 150 seconds. The router exported the resulting flow records to a collector which stored them in a database after anonymizing the IP addresses.

Our evaluation is not based on individual flows but on time-series of the number of bytes, packets, and flows counted in equally spaced time intervals. Each flow record was associated with the time interval in which the first packet passed the router. The interval length was set to 300 seconds (i.e., twice the flow timeout) in order to reduce possible distortions in the byte and packet counts that may result from long-lasting high-volume flows which are reported at the period of the active timeout. The flow count was determined as the number of distinct IP-five-tuples (i.e., cardinality of combinations of protocol, source and destination IP addresses and port numbers) to prevent manifold counting of flows reported in more than one records per time-interval.

We implemented the forecasting techniques and control charts in GNU Octave [11]. This approach enabled us to analyze the time-series data with different forecasting methods and control charts. For online traffic analysis, the detection mechanisms can be integrated into a real-time system, for example as a detection module of our traffic analysis framework TOPAS [17]. We determined the reason for the detected anomalies by identifying the responsible flows. Furthermore, we assessed the importance and relevance of the alarms for the network operator.

In the following subsections, we present the results for time-series of overall IP traffic, ICMP traffic, and SMB (Server Message Block) traffic. The objective is to answer the following questions:

- Which forecasting method is the most appropriate to generate residual time-series?
- Which control chart provides the best detection results when applied to these residuals?
- In which part of the traffic and in which metric do we find the most interesting anomalies?

We do not aim at finding the optimal solution, which is difficult regarding the numerous degrees of freedom. Also, the result would be limited to the specific set of measurement data. Instead, we are interested in recommendations that allow achieving good results under various conditions.

4.1. ANOMALY DETECTION IN OVERALL IP TRAFFIC

Figure 1 depicts the time-series of the total number of bytes, packets, and flows in the measurement period. All three metrics show a daily cycle of low values at nighttime and high values at daytime. Furthermore, we observe a weekly cycle with higher traffic on weekdays and lower traffic at weekends. October 3 is a public holiday in Germany, which results in slightly decreased traffic

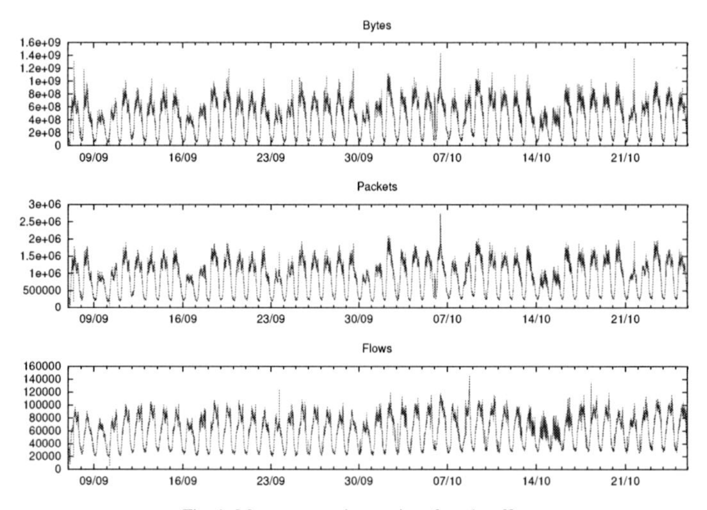

Fig. 1: Measurement time-series of total traffic

volume on this day as well. The regular run of the curves is interrupted by isolated peaks which are obvious traffic anomalies. Most of the time, a peak in one metric coincides with a peak in another metric. Yet, we rarely observe extreme values in all three metrics simultaneously.

In order to cope with the seasonal variation, we apply exponential smoothing and Holt-Winters forecasting and use the prediction errors as residual time-series. Given the measurement time-series y_t, we initialize the Holt-Winters components as follows:

$$L_s = 0 \quad ; \quad T_s = 0 \quad ; \quad I_i = y_i \text{ for } i = 1, \ldots, s$$

The seasonal period is set to $s = 288$ or $s = 2016$ to account for daily or weekly seasonality. With exponential smoothing, we obtain the first prediction value (and residual) in the second time interval ($t = 2$). In contrast, Holt-Winters forecasting requires the first s values for initialization, thus the first residual is generated at $t = s + 1$. To get comparable results, we only count the alarms raised after time interval $t = s$.

Figure 2 shows the residual time-series of byte counts for three different configurations of exponential smoothing ($\alpha = 1$, $\alpha = 0.5$, and $\alpha = 0.1$) and one setup of Holt-Winters forecasting with additive seasonal component ($s = 288$, $\alpha = 0.1$, $\beta = 0.001$, $\gamma = 0.25$). In the case of $\alpha = 1$, the residuals are simply the differences of consecutive measurement time-series values. Except for exponential smoothing with $\alpha = 0.1$, the seasonal variation of the mean is successfully removed. However, the variability of the residuals still depends on the time of day. Regarding the different settings for exponential smoothing, $\alpha = 0.5$ provides the best results: obvious anomalies in the original data appear as peaks, whereas the variability during normal traffic is relatively low. This visual impression is confirmed by the mean squared prediction error, which is the smallest for this setting.

Similar to our examinations of exponential smoothing, we tested various parameterizations of Holt-Winters forecasting with different smoothing constants, seasonal periods of one day and one week, additive and multiplicative seasonal components. The setting shown in Figure 2 effectively

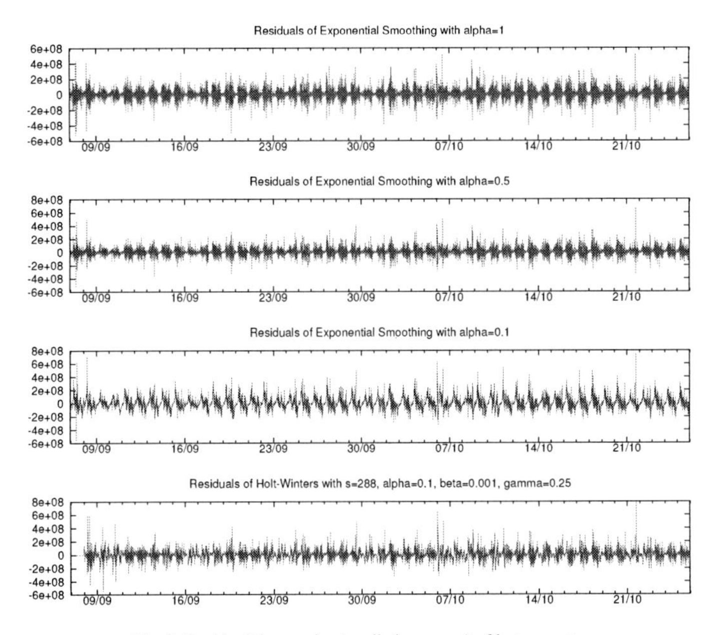

Fig. 2: Residual time-series (prediction errors) of byte counts

reduces the seasonal variation and exposes various anomalies in the measurement data. Yet, the additional complexity of Holt-Winters forecasting does not seem to ensure better results than simple exponential smoothing: the residual time-series of the two methods turned out to be quite similar. A possible explanation is that the seasonal period is very long (288 or 2016 intervals), hence the effect of the seasonal variation on consecutive values is very small.

Figure 3 shows the sample autocorrelation of the original byte count time-series and the corresponding residuals. As expected, the seasonality of the original measurements reappears in the autocorrelation plot. On the other hand, the serial correlation in the residual time-series attenuates quite quickly.

We applied the Shewhart control chart of individuals, the two-sided CUSUM control chart, and the EWMA control chart to the residual time-series. Control limits were defined as multiples of $\hat{\sigma}$ which was estimated by EWMS (see Section 3.3). The smoothing constant ρ controls how quickly the limits adapt to variability changes in the prediction errors. For our purposes, $\rho = 0.01$ turned out to be a good setting.

Figure 4 shows the measurement time-series of byte counts on top and three control charts applied to the residuals of exponential smoothing below. The parameters of the control charts are as follows:

- Shewhart: $UCL = -LCL = 6\hat{\sigma}$
- CUSUM: $K = \hat{\sigma}$; $h = 6\hat{\sigma}$
- EWMA: $\lambda = 0.25$; $UCL = -LCL = 5\hat{\sigma}\sqrt{\frac{\lambda}{2-\lambda}}$

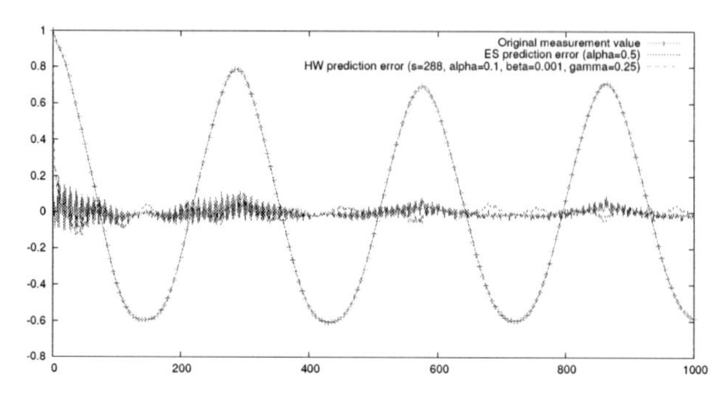

Fig. 3: Sample autocorrelation of byte counts

Table 1: Byte Anomalies Detected in All Control Charts

Day	Cause
08/09	FTP download (640 MBytes at 1 MByte/s on average)
24/09	RSF-1 data transfer (approx. 310 MBytes)
29/09	HTTP download (approx. 160 MBytes)
06/10	High SMTP traffic during 3 hours in the night
11/10	HTTP download (peak of 355 MBytes in one interval)
18/10	HTTP download (peak of 195 MBytes in one interval)
21/10	HTTP download (peak of 524 MBytes in one interval)

The Shewhart control chart shows the residual time-series and the corresponding control limits. The CUSUM control chart depicts the maximum of the two CUSUM statistics g_t^+ and g_t^- as well as the threshold h. The EWMA chart finally shows the exponentially smoothed residuals z_t and the control limits.

The dotted vertical lines in Figure 4 mark the intervals in which the corresponding value is beyond the control limits. We obtained 11, 15, and 11 alarms for Shewhart, CUSUM, and EWMA, respectively. Some of them are so close to each other that they can hardly be distinguished in the figure. Table 1 lists the set of anomalies that are detected in all three charts. For each anomaly, we identified the responsible flows and found that most of the alarms were caused by large downloads from web or file servers. What we describe as RSF-1 data transfer in the table is a large flow to UDP port 1195, which has been registered by High-Availability.Com [13] for a high-availability and cluster middleware application. Very probably, these downloads represent legitimate traffic. However, we detected anomalous high SMTP traffic on October 6 lasting for several hours, which is a sign of a mailbomb triggered by spammers or a worm propagating via e-mail. Most of the remaining alarms not mentioned in the table could be explained by the same kinds of HTTP, FTP, and RSF-1 traffic. Though, some of the alarms trigged by the CUSUM and EWMA control charts could not be associated to any unusual pattern in the flow records.

Some of the detected anomalies also appear as extreme values in the original measurement data. Hence, they could be detected with a threshold applied to the byte counts directly. Others, such as the mailbomb, do not cause extraordinarily high byte counts, i.e. they can only be detected in the residuals.

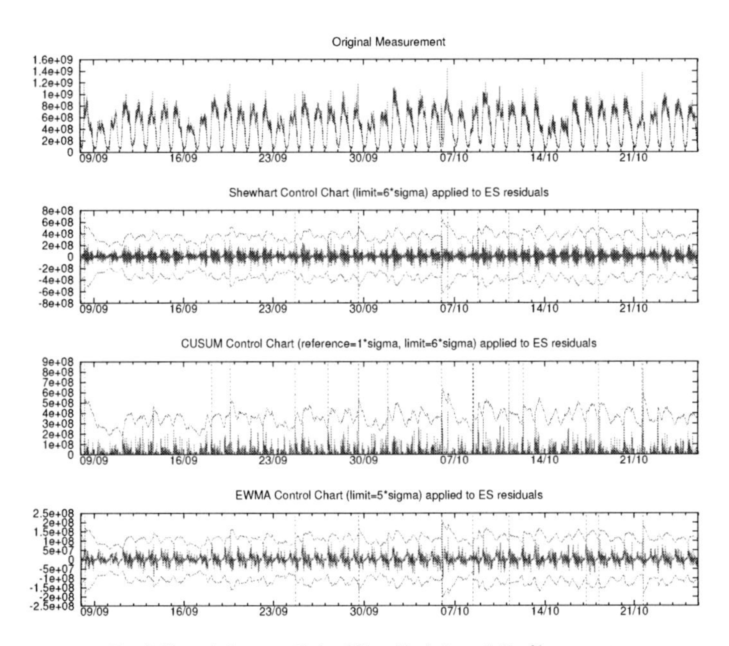

Fig. 4: Control charts applied to ES residuals ($\alpha = 0.5$) of byte counts

We applied the same control charts to the Holt-Winters residuals and obtained similar results as for exponential smoothing. Furthermore, we examined if more interesting anomalies can be found in the packet and flow counts or in any ratio of the three basic metrics, such as the average number of bytes per flow. Packet and byte counts triggered the same alarms. A couple of new anomalies were found in the flow counts. One of these alarms is the result of a large number of short SSH connections from one client to multiple servers, a pattern that may be caused by a massive password guessing attempt. Another alarm coincides with a time interval in which the traffic abruptly breaks down, possibly due to a network failure. Regarding the anomalies found in the ratios, we did not notice any improvements compared to the basic metrics of bytes, packets, and flows.

As a result, we conclude that residual generation using exponential smoothing techniques and change detection with the Shewhart control chart of individuals enables the detection of traffic anomalies with relatively low computational complexity. The CUSUM and EWMA control chart did not provide better detection results but raised additional alarms that could not be linked to anomalous traffic behavior. In the EWMA control chart, the moving average flattens short peaks in the residuals and thus hampers their detection. However, such peaks result from abrupt changes in the original measurement data, which are events we definitively want to detect.

An appropriate level of the control limits needs to be determined by experimentation in order to focus on the most significant anomalies. Among the detected anomalies in the overall traffic, the mail-bomb, the password guessing attempt, and the network failure are the most interesting events for the network operator. However, the majority of the alarms is caused by legitimate traffic, independently of the considered metric.

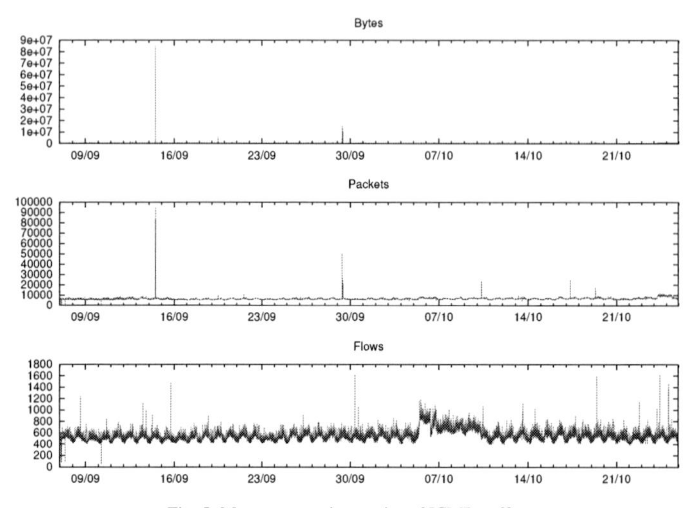

Fig. 5: Measurement time-series of ICMP traffic

4.2. ANALYZING ICMP TRAFFIC

As part of the Internet protocol suite, the Internet Control Message Protocol (ICMP) is mainly used for exchanging error messages, for example, if a certain host cannot be reached due to link or routing problems. ICMP is also used for network testing and debugging purposes (e.g., using ping and traceroute commands) and self configuration in local IP networks. As ICMP is not directly involved in the transport of user and application data, we expect a low and time-invariant level of ICMP traffic under normal conditions. Indeed, we can observe this behavior in the byte and packet time-series shown in Figure 5. In contrast, the number of flows shows daily variation, yet less pronounced than in the overall traffic.

We adopted the most promising approach from Section 4.1, namely the Shewhart control chart applied to the residuals of exponential smoothing ($\alpha = 0.5$), to detect anomalies in the ICMP traffic. Control limits at $\pm 6\hat{\sigma}$, as used before, generated a very large number of alarms for byte and packet counts. Therefore, we increased the control limits to $\pm 8\hat{\sigma}$ in order to focus on the most significant anomalies.

The anomalies found in the byte and packet time-series are listed in Table 2. Many alarms are triggered by both metrics, especially those caused by ping traffic (ICMP echo requests and replies). Sporadic occurrences of ping traffic at moderate rate are not suspicious, hence the corresponding alarms are not of much interest. The extremely high impulse on September 14 is the result of one host pinging another host at very high rate, which can be a sign of an attack. Though, as the ping is very short, we think that it was executed for debugging purposes. Apart from ping traffic, many traffic anomalies are caused by destination unreachable messages, most of them reporting that a large packet could not be fragmented due to the 'don't fragment' bit in the IP header. The corresponding ICMP messages are quite large because they include a section of the dropped packet. Therefore, these anomalies are mainly detected in the number of bytes.

Table 2: Shewhart Alarms for ICMP Traffic: Bytes and Packets

Time	Bytes alarm	Packets alarm	Cause
11/09 09:00	x		Echo replies from/to one host
12/09 15:15	x		Ping (moderate rate)
14/09 13:50	x	x	Ping flood (very high rate)
14/09 14:25	x		Ping flood (very high rate)
14/09 14:40	x		Ping flood (very high rate)
19/09 14:00	x	x	Echo replies from/to one host
21/09 15:30	x	x	Destination unreachable (fragmentation required)
24/09 08:25	x		Destination unreachable (fragmentation required)
29/09 10:40	x	x	Ping (moderate rate)
29/09 11:15	x		Ping (moderate rate)
29/09 11:30	x		Ping (moderate rate)
05/10 16:35	x		Destination unreachable (fragmentation required)
06/10 18:00	x		Destination unreachable (fragmentation required)
10/10 10:00	x	x	Time exceeded from one host
13/10 07:50		x	Destination port unreachable
16/10 16:45	x		Destination unreachable (fragmentation required)
17/10 09:55	x	x	Ping (moderate rate)
19/10 09:30	x	x	Ping (moderate rate)
21/10 23:00	x		Destination unreachable (fragmentation required)
25/10 15:35	x		Ping (moderate rate)

Table 3: Shewhart Alarms for ICMP Traffic: Flows

Time	Cause
13/09 14:45	Destination port unreachable from many sources to one host
15/09 19:40	ICMP scan followed by TCP connections to port 3389 (WBT)
30/09 11:05	ICMP scan followed by TCP connections to port 1433 (MS MSQL)
19/10 12:20	ICMP scan followed by TCP connections to ports 3389 (WBT) and 1433 (MS SQL)
24/10 13:20	Ping at moderate rate to five hosts
25/10 05:40	ICMP scan followed by TCP connections to ports 80 and 3128 (HTTP proxies)

As can be seen in Table 3, the flow count residuals exceed the control limits only six times. None of these alarms coincides with any of the byte and packet alarms. Examining the flow records, four of the alarms can be explained by ICMP echo requests sent by individual hosts to a few hundred IP addresses. Echo replies are returned from a small proportion of the scanned IP addresses only. To these destinations, the scanning host then tries to establish TCP connections on ports 3389, 1433, 80, or 3128 which are used by Microsoft remote desktop (Windows-based Terminal, WBT), Microsoft SQL server, and HTTP proxies, respectively. ICMP scans are typically performed with help of automated network scanners in order to detect active hosts. It is difficult to assess if the observed traffic is harmful or not. Maybe the scans served testing and debugging purposes. This assumption is fortified by our experience that malware and worms usually try to establish TCP connection directly without preceding ICMP scans. However, it has been recently reported that ICMP scans are more and more frequently deployed in advance of an infection attempt [10] as well.

Having a look at Figure 5, we see that the number of flows is increased between October 5 and October 10. After decreasing the Shewhart control limits to $\pm 5\hat{\sigma}$, this anomaly is also detected in the flow count residuals. We examined the flow records and discovered that the increase is caused by ICMP destination unreachable messages sent from different sources to one specific host. Two different error codes are reported: host unreachable and communication administratively prohibited. The second one is returned by routers or firewalls if a packet is discarded because of a blocked destination IP address or port number. The host receiving all these messages thus had to be emitting a lot of packets to non-existing or blocked destinations. Indeed, we found a lot of outgoing connection requests from this IP address to TCP ports 445 (Microsoft-DS) and 139 (Netbios) during five days. On Microsoft Windows systems, these ports are well-known for many vulnerabilities which are exploited by worms.

All in all, anomalies found in the ICMP traffic give the network operator valuable insights in the current state of the network. An anomalous increase of the number of destination unreachable messages indicates a network failure or the occurrence of a TCP or UDP scan performed by worms or hackers. Large numbers of ICMP flows are mostly caused by ICMP scans which do not represent an instantaneous security threat but often reveal other suspicious traffic, such as connection attempts to specific TCP ports following a scan.

4.3. ANALYZING SMB TRAFFIC

Motivated by the findings in the ICMP traffic, we analyzed TCP traffic to and from port 445. Since Windows 2000, this port is used by Microsoft for file and printer sharing in local area networks via the SMB (Server Message Block) protocol. However, vulnerabilities in this service are also being exploited by worms to infect unprotected computers in the network. A prominent example is the Sasser Worm which has been spreading over the Internet since 2004.

For our analysis, we consider the difference of TCP traffic to and from port 445. The plots in Figure 6 show the corresponding time-series for the number of bytes, packets, and flows. As can be seen, all metrics have small values close to zero most of the time and do not show any seasonal variation. Between October 5 and 9, we observe longer periods of large positive values, which means that many more bytes, packets, and flows are directed to port 445 than returned. During this time, we also observed an increased number of ICMP destination unreachable messages. Indeed, the anomalies in the ICMP and SMB traffic are related to each other: the emitter of the SMB traffic is the receiver of the ICMP traffic. As mentioned in Section 4.2, the host is very probably infected by a worm trying to connect to randomly chosen destinations.

As before, we applied Shewhart control charts to the prediction errors of exponential smoothing. We set the control limits to $\pm 10\hat{\sigma}$ in order to get a reasonably small number of alarms. We obtained 23 alarms for the byte count residuals, 12 alarms for the packet count residuals, and 13 alarms for the flow count residuals. While many of the byte and packet alarms are caused by non-suspicious SMB traffic (e.g., data transfers between two hosts), all of the flow alarms are triggered by scanning activities. Four of the flow alarms are related to the worm-infected host already mentioned, the remaining alarms are caused by short scans originating from different IP addresses. These scans probably belong to worm traffic generated in distant networks, thus only parts of it are observed by the router.

4.4. DISCUSSION OF RESULTS

Our evaluation demonstrates the applicability of forecasting techniques and control charts for detecting traffic anomalies in time-series of byte, packet, and flow counts. The prediction error of

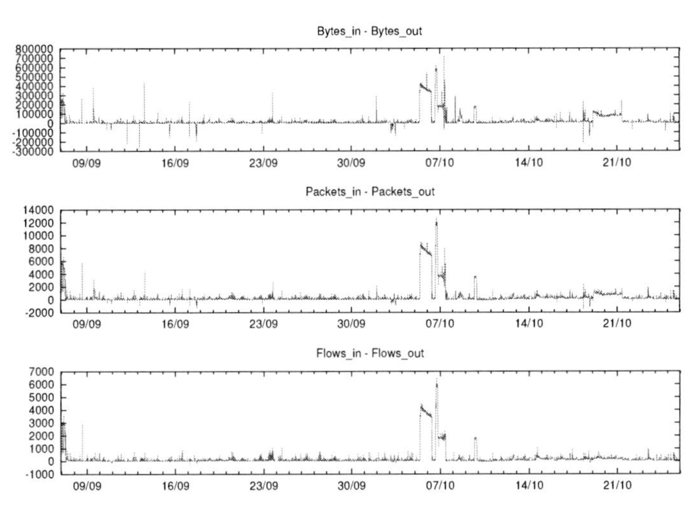

Fig. 6: Measurement time-series of SMB traffic

exponential smoothing with smoothing constant $\alpha = 0.5$ turned out to be a robust residual generation method which provides good results for various traffic metrics. Among the examined change detection mechanisms, the Shewhart control chart of individuals works fine despite of its simplicity. The lack of knowledge of the residuals' distribution under normal conditions inhibits the calculation of exact control limits for a given false alarm level. Yet, the sensitivity of the detection can be very easily adjusted by defining empirical control limits as multiples of the estimated standard deviation.

From a theoretical point of view, CUSUM and EWMA control charts are better in detecting small sustained shifts in the mean. However, the forecasting-based residual generation is characterized by a differentiation effect: abrupt changes in the measurement time-series result in short impulses in the prediction errors. Therefore, sustained shifts rarely occur in the residual time-series monitored in the control charts.

The relevance of the detected anomalies depends very much on the analyzed traffic and the considered metrics. Most byte and packet anomalies detected in the overall traffic as well as in the SMB traffic were caused by large data transfers. Among these uninteresting alarms, events of actual importance risk to go unnoticed. Therefore, it is advisable to monitor traffic metrics that are less influenced by unpredictable but legitimate behavior of users and applications. Examples are the numbers of ICMP and SMB flows as well as the number of ICMP destination unreachable messages. where most anomalies are caused by suspicious traffic.

In our approach, control limits are calculated relatively to the EWMS estimation of the standard deviation. The benefit of this approach is that the limits dynamically adapt to changes in the residuals. However, we do not stop the update of the limits if an anomaly is detected. Therefore, the control limits are often increased to very high values after an alarm, as can be observed in Figure 4. This problem could be solved by temporarily suspending the update of the EWMS estimator after the detection of an anomaly.

5. RELATED WORK

Hood and Ji [14] convert MIB variables into a measurement time-series, eliminate serial correlation by fitting an AR(2) model, and detect network failures in the residuals. Hellerstein et al. use the GLR (generalized likelihood ratio) algorithm to detect anomalies in the number of web server requests per five minutes interval [12]. Systematic changes are eliminated by estimating daily and weekly variations as well as monthly trend from a set of training data. In addition, an AR(2) model is fitted to remove the remaining serial correlation. Brutlag [6] employs Holt-Winters forecasting to model baseline, trend, and daily variation in the outgoing traffic of a web server. Barford et al. [1] apply Holt-Winters forecasting to time-series of packet, byte, and flow counts as a reference anomaly detection approach for their own detection mechanism based on wavelets. The evaluation yields similar detection performance for the two approaches.

Ye et al. use EWMA control charts to detect anomalies in computer audit data [32]. The results are compared to those obtained with a Shewhart individuals control chart applied to the prediction errors of exponential smoothing†. Like in our work, the control limits depend on the EWMS estimate of the standard deviation. Paul [19] adopts this method for detecting denial-of-service attacks against web servers.

The optimality of the CUSUM algorithm [5] is frequently brought up to justify its usage for traffic anomaly detection. For example, Wang et al. deploy the CUSUM algorithm to detect SYN flooding attacks. The considered metrics are calculated from the number of TCP SYN, FIN, and SYN/ACK packets [30, 31]. Peng et al. [20] apply the CUSUM algorithm to the number of RST packets returned in response to SYN/ACK packets in order to detect reflector attacks. In [21], the same authors count the number of new source IP addresses to detect distributed denial-of-service attacks. Siris and Papagalou [27] use exponential smoothing to generate prediction errors for the number of SYN packets. The residuals serve as input to CUSUM in order to detect SYN flooding attacks. Similarly, Rebahi and Sisalem [22] use the number of SIP (Session Initiation Protocol) INVITE messages to detect denial-of-service attacks against SIP servers. In order to be optimal, the CUSUM algorithm must be applied to a time-series of independent observations belonging to a specific family of probability distributions. However, none of these works shows that these conditions are fulfilled, hence it is unsure if the CUSUM algorithm actually is the best choice.

The research group of Tartakovsky has proposed several approaches to apply the CUSUM control chart to multivariate data. In [3] and [29], they calculate a chi-square statistic as input for CUSUM in order to detect denial-of-service attacks. For the same purpose, the multichart CUSUM algorithm proposed in [15] and [28] performs separate tests on each component of the multivariate data. Salem et al. apply the multichart CUSUM algorithm to the entries of a count-min sketch to detect SYN flooding attacks and scans [25]. Common to these multivariate methods is the assumption that the components in the multivariate data are mutually independent, which is usually not fulfilled in the case of traffic measurement data. Tartakovsky at al. also downplay the prerequisite of uncorrelated observations arguing that the false alarm rate decays exponentially fast for increasing thresholds [28] under conditions that are to be usually satisfied. Yet, they do not verify if these conditions are fulfilled by the data used in their evaluation.

†The authors misleadingly call this approach "EWMA control chart for autocorrelated data" although it actually is a Shewhart control chart.

6. CONCLUSION

We evaluated the applicability of control charts for detecting traffic anomalies. A necessary requirement is the removal of systematic changes and serial correlation from the measurement time-series. We showed that both, seasonal variation and serial correlation can be effectively reduced with robust forecasting techniques based on exponential smoothing. Comparing three different control charts, we determined that CUSUM, although favored by many related works, does not perform better than the simpler Shewhart control chart of individuals when applied to time-series of prediction errors.

Our evaluation based on traffic measurement data collected in an ISP backbone network shows that many anomalies are provoked by legitimate traffic. To increase or decrease the total number of alarms, it suffices to adjust the control limits. Yet, according to our experience, the proportion of alarms that are relevant for the network operator mainly depends on the monitored metrics and the parts of traffic analyzed.

In order to validate our findings, we will conduct similar experiments with measurement data obtained in other networks. Moreover, it will be interesting to examine if dependencies between different metrics can be exploited in a multivariate residual generation process. We observed strong correlation of the number of bytes, packets, and flows in the traffic measurement data, so the detection of changes in the correlation structure may allow us to discover anomalies which cannot be detected in a single metric.

REFERENCES

[1] Barford, P., Kline, J., Plonka, D., and Ron, A. A signal analysis of network traffic anomalies. In *ACM SIGCOMM Internet Measurement Workshop 2002* (Marseille, France, Nov. 2002).

[2] Basseville, M., and Nikiforov, I. V. *Detection of abrupt changes: Theory and application*. Prentice-Hall, Inc, 1993.

[3] Blazek, R. B., Kim, H., Rozovskii, B. L., and Tartakovsky, A. G. A novel approach to detection of denial-of-service attacks via adaptive sequential and batch-sequential change-point detection methods. In *IEEE Workshop on Information Assurance and Security* (West Point, NY, USA, June 2001).

[4] Box, G. E. P., and Jenkins, G. M. *Time-Series Analysis, Forecasting and Control*, 2 ed. Holden-Day, San Francisco, USA, 1970.

[5] Brodsky, B., and Darkhovsky, B. *Nonparametric Methods in Change-Point Problems*, vol. 243 of *Mathematics and its applications*. Kluwer Academic Publishers, 1993.

[6] Brutlag, J. D. Aberrant behavior detection in time series for network monitoring. In *14th Systems Administration Conference (LISA 2000)* (New Orleans, Louisiana, USA, Dec. 2000), U. Association, Ed.

[7] Chatfield, C. *The analysis of time series: an introduction*, 6 ed. CRC Press LLC, 2003.

[8] Claise, B., Bryant, S., Sadasivan, G., Leinen, S., Dietz, T., and Trammell, B. H. Specification of the IP Flow Information Export (IPFIX) Protocol for the Exchange of IP Traffic Flow Information. RFC 5101 (Proposed Standard), Jan. 2008.

[9] Claise, B., Sadasivan, G., Valluri, V., and Djernaes, M. Cisco Systems NetFlow Services Export Version 9. RFC 3954 (Informational), Oct. 2004.

[10] Dacier, M. Leurré.com: a worldwide distributed honeynet, lessons learned after 4 years of existence. In *Presentation at Terena Networking Conference* (Bruges, Belgium, May 2008).

[11] GNU OCTAVE HOMEPAGE. http://www.octave.org, 2008.

[12] HELLERSTEIN, J. L., ZHANG, F., AND SHAHABUDDIN, P. Characterizing normal operation of a web server: Application to workload forecasting and capacity planning. In *24th International Computer Measurement Group (CMG) Conference* (Anaheim, California, USA, Dec. 1998).

[13] HIGH-AVAILABILITY.COM HOMEPAGE. http://www.high-availability.com, 2008.

[14] HOOD, C. S., AND JI, C. Proactive network fault detection. In *IEEE Conference on Computer Communications (INFOCOM'97)* (Kobe, Japan, Apr. 1997), pp. 147–1155.

[15] KIM, H., ROZOVSKII, B. L., AND TARTAKOVSKY, A. G. A nonparametric multichart cusum test for rapid detection of dos attacks in computer networks. *International Journal of Computing & Information Sciences 2*, 3 (Dec. 2004), 149–158.

[16] MONTGOMERY, D. C. *Introduction to Statistical Quality Control*, 5 ed. John Wiley & Sons, 2005.

[17] MÜNZ, G., AND CARLE, G. Real-time Analysis of Flow Data for Network Attack Detection. In *Proc. of IFIP/IEEE Symposium on Integrated Management (IM 2007)* (Munich, Germany, May 2007).

[18] PAGE, E. Continuous inspection schemes. *Biometrika 41* (1954), 100–115.

[19] PAUL, O. Improving web servers focused dos attacks detection. In *IEEE/IST Workshop on Monitoring, Attack Detection and Mitigation (MonAM 2006)* (Tübingen, Germany, Sept. 2006).

[20] PENG, T., LECKIE, C., AND RAMAMOHANARAO, K. Detecting reflector attacks by sharing beliefs. In *IEEE 2003 Global Communications Conference (Globecom 2003)* (2003).

[21] PENG, T., LECKIE, C., AND RAMAMOHANARAO, K. Proactively detecting distributed denial of service attacks using source ip address monitoring. In *Networking 2004* (Athens, Greece, May 2004).

[22] REBAHI, Y., AND SISALEM, D. Change-point detection for voice over ip denial of service attacks. In *15. ITG/GI Fachtagung Kommunikation in Verteilten Systemen (KiVS)* (Bern, Switzerland, Feb. 2007).

[23] ROBERTS, S. Control chart tests based on geometric moving averages. *Technometrics 1* (1959), 239–250.

[24] ROBINSON, P., AND HO, T. Average run lengths of geometric moving average charts by numerical methods. *Technometrics 20* (1978), 85–93.

[25] SALEM, O., VATON, S., AND GRAVEY, A. An efficient online anomalies detection mechanism for high-speed networks. In *IEEE Workshop on Monitoring, Attack Detection and Mitigation (MonAM2007)* (Toulouse, France, Nov. 2007).

[26] SHEWHART, W. *Economic Control of Quality Manufactured Product*. D.Van Nostrand Reinhold, Princeton, NJ, 1931.

[27] SIRIS, V. A., AND PAPAGALOU, F. Application of anomaly detection algorithms for detecting syn flooding attacks. In *IEEE Global Telecommunications Conference (Globecom 2004)* (Dallas, USA, Nov. 2004).

[28] TARTAKOVSKY, A. G., ROZOVSKII, B. L., BLAZEK, R. B., AND KIM, H. Detection of intrusions in information systems by sequential change-point methods. *Statistical Methodology 3*, 3 (2006).

[29] TARTAKOVSKY, A. G., ROZOVSKII, B. L., BLAZEK, R. B., AND KIM, H. A novel approach to detection of intrusions in computer networks via adaptive sequential and batch-sequential change-point detection methods. *IEEE Transactions on Signal Processing 54*, 9 (Sept. 2006).

[30] WANG, H., ZHANG, D., AND SHIN, K. G. Detecting syn flooding attacks. In *IEEE Infocom 2002* (June 2002).

[31] WANG, H., ZHANG, D., AND SHIN, K. G. Syn-dog: Sniffing syn flooding sources. In *22nd International Conference on Distributed Computing Systems (ICDCS'02)* (Vienna, Austria, July 2002).

[32] YE, N., VILBERT, S., AND CHEN, Q. Computer intrusion detection through ewma for autocorrelated und uncorrelated data. *IEEE Transactions on Reliability 52*, 1 (Mar. 2003), 75–82.

Martin BURKHART*, Daniela BRAUCKHOFF†, Martin MAY‡,

On the Utility of Anonymized Flow Traces for Anomaly Detection

The sharing of network traces is an important prerequisite for the development and evaluation of efficient anomaly detection mechanisms. Unfortunately, privacy concerns and data protection laws prevent network operators from sharing these data. Anonymization is a promising solution in this context; however, it is unclear if the sanitization of data preserves the traffic characteristics or introduces artifacts that may falsify traffic analysis results. In this paper, we examine the utility of anonymized flow traces for anomaly detection. We quantitatively evaluate the impact of IP address anonymization, namely variations of permutation and truncation, on the detectability of large-scale anomalies. Specifically, we analyze three weeks of un-sampled and non-anonymized network traces from a medium-sized backbone network. We find that all anonymization techniques, except prefix-preserving permutation, degrade the utility of data for anomaly detection. We show that the degree of degradation depends to a large extent on the nature and mix of anomalies present in a trace. Moreover, we present a case study that illustrates how traffic characteristics of individual hosts are distorted by anonymization.

1. Introduction

One of the principal reasons for the slow progress in anomaly detection research is the lack of publicly available, unaltered network traffic traces. The sharing of traffic data is hindered since releasing data always introduces a threat to users' privacy. Even when data export is restricted to packet headers, as is the case with Cisco NetFlow, a certain amount of personal information may still be extracted and exploited to illegitimately profile user behavior. This threat has already been recognized by data protection legislation in both Europe [6, 7] and the United States [15]. As a result, multiple anonymization tools that prevent the leakage of privacy information have been developed, such as FLAIM [18], TCPdpriv [13], and CryptoPAn [8].

For practical reasons these tools have been evaluated only with regard to privacy concerns (e.g., in [4, 10, 3, 17]). The remaining question is whether the data sanitization preserves the traffic characteristics, or introduces artifacts that compromise the utility for research purposes. For researchers and engineers with access to unanonymized data sets, this is not an issue; unfortunately, only few research institutes have such traffic traces available and the large majority works with publicly available, but already anonymized data sets. As a result, a study on the impact of anonymization methods

*Computer Engineering and Networks Laboratory, ETH Zurich, Switzerland, burkhart@tik.ee.ethz.ch
†brauckhoff@tik.ee.ethz.ch
‡may@tik.ee.ethz.ch

is essential and indeed overdue, since numerous anomaly detection algorithms have been evaluated with anonymized data. Therefore, the goal of this work is to determine to what extent the anonymized traces falsify the results of commonly used anomaly detection mechanisms.

To the best of our knowledge, the specific problem we are investigating has not yet been addressed in literature. In [19], Soule et al. study NetFlow data from two backbone networks that apply different sampling and anonymization schemes, and suggest that anonymization might have an impact on anomaly detection. The problem of data loss due to anonymization is also identified in [25], where the authors give qualitative recommendations for anonymization of NetFlow logs when security services are outsourced. Yurcik et al. [24] analyze single-field anonymization tradeoffs with regard to intrusion detection. Unfortunately, their dataset contains already anonymized IP addresses, hence the impact of IP address anonymization techniques on utility is not studied. On the contrary, [2] and [12] studied the utility of sampled traffic traces without addressing the effect of anonymization.

This paper represents the first comprehensive study on the utility of anonymized data for statistical anomaly detection approaches. Specifically, we focus on the anonymization of IP addresses as they comprise the biggest threat to user privacy. We analyze the most popular IP address anonymization techniques, namely blackmarking, truncation, random permutation, and (partial) prefix-preserving permutation. Our contributions are as follows: (i) we introduce a generic methodology for evaluating the impact of anonymization on flow-based traffic analysis applications; (ii) we quantify the utility of anonymized data for backbone anomaly detection with the help of a three-week long data set from a medium-size ISP and an anomaly detector based on a Kalman filter; and (iii) we present an overall estimate for the impact of anonymization on the analysis of individual hosts' traffic characteristics.

To tackle the problem of determining the utility of anonymized data, we first introduce the granularity design space for traffic analysis. This design space has two dimensions: the *subset size* of the address space under investigation and the *resolution* of the examination. We argue that the complete granularity design space is valuable for traffic analysis, even though today, only a subset of it is used (e.g., Origin-Destination flows). We then show how the granularity design space is diminished by the considered anonymization techniques.

In section 3 and 4, we present a measurement study on the impact of anonymization on the data utility for anomaly detection. We introduce the Kalman filter used for anomaly detection and evaluate the detectors' performance on the unanonymized and unsampled, manually labeled three-week data set. We assess the utility of anonymized data by evaluating the detector performance on the restricted set of metrics available with each anonymization scheme. Specifically, we evaluate the results with the help of ROC curves [9] that plot false positives vs. true positives for a range of thresholds. We show that the restriction of available granularities through anonymization degrades the performance of anomaly detection. Surprisingly, we find that the degree of degradation differs for volume, scan/DoS, and network fluctuation anomalies, as well as for UDP and TCP traffic. As an example, we will show that for the detection of network fluctuations, the utility drops from 87% to 70% when the number of truncated bits is doubled.

In section 5, we study the effect of anonymization on detailed traffic analysis and root cause identification. We are able to show that even a small restriction of the subset size heavily impacts the visibility of anomalies. Finally we discuss our findings and conclude the paper in section 6.

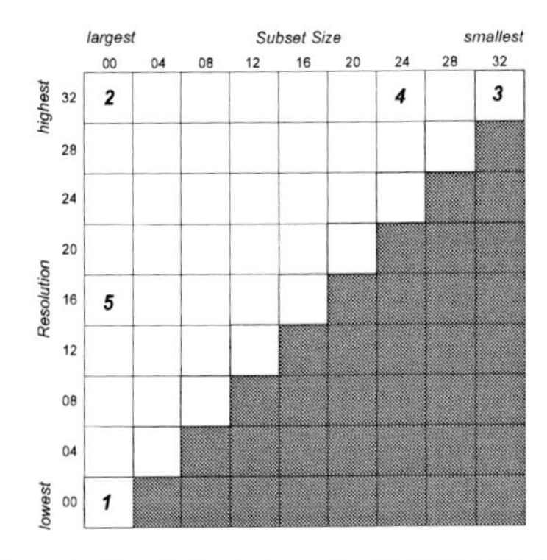

Figure 1: Granularity design space for metrics used in statistical anomaly detection.

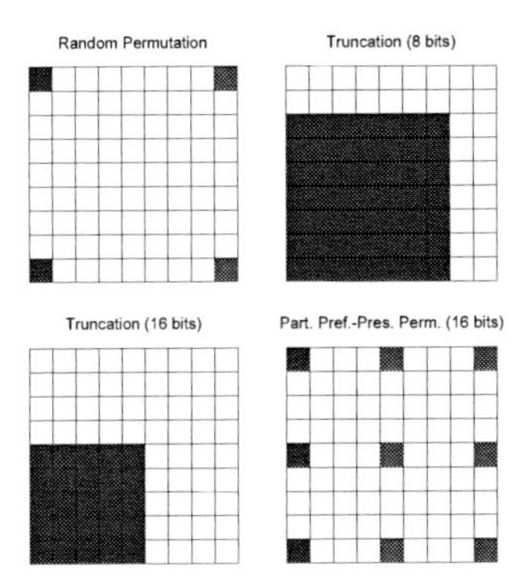

Figure 2: Resolutions and subset sizes available with different anonymization techniques.

2. Utility of Anonymized Data for Anomaly Detection

In this section, we investigate how five popular IP address anonymization techniques impact statistical anomaly detection on flow data. To make our discussion more systematic, we introduce a granularity design space for traffic analysis, and show how the different anonymization techniques diminish this design space.

2.1. A Granularity Design Space

Up to now the metrics used for anomaly detection, and traffic analysis in general, have been designed in an ad-hoc manner, based on (i) the characteristics of the data set under study, and (ii) the type of traffic characteristic one is interested in. Prominent examples of such metrics are the well-known volume metrics, such as byte, packet, and flow counts, which are simply computed over all traffic in a given trace. Lakhina et al. in [11] group the *anonymized* traffic from the Abilene network into Origin-Destination (OD) flows before analyzing it further with Principal Component Analysis. Whereas, in [22, 23] host-based metrics such as IP address entropy or the number of active connections are used for host profiling in a clustering algorithm.

To investigate the impact of anonymization in a systematic manner, however, it is necessary to explore the whole granularity design space for traffic analysis. To motivate the design space it is helpful to use an allegory from image processing or photography. When taking a picture, one focuses on the object of interest and selects a fine or coarse resolution to display the details at the desired level. Similarly, the granularity design space has two dimensions:

Subset size The size of the network that is to be analyzed. When analyzing backbone data for example, we can analyze all traffic, or we can focus on a specific subnet in the backbone.

Resolution The address granularity at which the traffic is analyzed. We can select a very high resolution of IP addresses if one is interested in profiling hosts, or a low resolution of Autonomous Systems (or OD flows) if one is interested in more global events.

We give a matrix representation of the granularity design space in Fig. 1. The x-axis ranges from the largest subset size of all traffic (00 bits), to the smallest possible subset size of a single IP address (32 bits). Likewise the y-axis ranges from the highest resolution of individual IP addresses (32 bits) to the lowest resolution of all traffic (00 bits). The design space matrix can be divided in three sections: the upper triangle, the diagonal, and the lower triangle. Traditional volume metrics fall in one of the cells on the diagonal, where the subset size equals the resolution (e.g., we select all traffic from a /16 subnet and compute the flow counts over all traffic coming from that /16 network). The upper triangle features metrics where the resolution level is smaller than the size of the selected traffic subset. Such metrics, e.g, the number of unique IP addresses in the inbound traffic to a selected /24 subnet, are also frequently used for traffic analysis. Finally, for metrics on the lower triangle, the resolution level is larger than the size of the selected traffic subset. Such metrics are rarely used today, and are thus of less interest to our study.

Note, the full design space is not always available. For example, when working with data from stub networks, e.g., a campus network, where the maximum available subset size equals the IP address range assigned to the studied network. For a /16 campus network the subset size may range from 16 to 32, i.e., subset sizes larger than 16 bit are not available. Nevertheless, to keep the subsequent discussion general we assume the whole design space is available.

To further illustrate the granularity design space, we give example metrics for five selected cells of the matrix:

- Cell 1 [00,00]: Select all traffic and set the resolution to the minimum. An example metric is the well-known volume over all traffic.
- Cell 2 [00,32]: Select all traffic and set the resolution to the maximum. Examples are the volume per IP address or the number of unique IP addresses in all traffic.
- Cell 3 [32,32]: Select traffic to/from one IP address and set the resolution to the maximum. Metrics falling in this category are, e.g., the number of unique ports per IP address, or the number of unique IP addresses that the host under observation sends traffic to.
- Cell 4 [24,32]: Select traffic to/from one /24 network and set the resolution to the maximum. Examples for this case are the flow count per IP address, or the unique number of IP addresses that send traffic in the monitored /24 network.
- Cell 5 [00,16]: Select all traffic and set the resolution to /16 networks. An example metric is the volume per /16 networks in all traffic.

2.2. How Anonymization Diminishes the Design Space

In the following we outline the studied IP address anonymization techniques and show how they diminish the granularity design space. The most commonly employed IP address anonymization techniques are blackmarking, truncation, random permutation, prefix-preserving permutation, and partial prefix-preserving permutation. An illustrative example for each technique, except blackmarking which is trivial, is given in Table 1.

The available subset of the design space for different anonymization techniques is illustrated in Fig. 2, where filled squares mark the possible combinations of subset size and resolution for each anonymization technique. Note that for permutation-based approaches all fields with a subset size smaller than 00 are marked with a different color. We did this to signify that subsets of smaller sizes

IP Address	Truncation (16 bits)	Random Permutation	Prefix-Pres. Permutation	Partial Prefix-Pres. Permutation (16 bits)
129.132.91.35	129.132.0.0	112.4.23.73	22.5.99.76	73.9.8.1
129.132.91.177	129.132.0.0	62.12.96.67	22.5.99.41	73.9.181.17
129.132.8.37	129.132.0.0	205.72.5.18	22.5.181.92	73.9.1.230
152.88.3.90	152.88.0.0	2.14.12.133	110.27.20.1	18.7.18.133
152.96.99.2	152.96.0.0	19.0.111.20	110.9.0.12	24.125.43.6
82.130.102.115	82.130.0.0	12.171.92.3	145.21.5.19	145.213.2.77

Table 1: Examples of IP address anonymization

may be distinguished, but not identified, since the mapping from real to anonymized IP addresses is usually not known. Hence, a subset of interest has to be identified by different means, e.g., by selecting subnets with particular traffic characteristics. For the subsequent analysis, however, we make no distinction between the two cases.

Blackmarking (BM) is the simplest of all studied anonymization techniques. It blindly replaces all IP addresses in a trace with the same value. As a result, all information about individual IP addresses or subnets is lost and only metrics with the lowest resolution and the largest subset size, e.g., the volume over all traffic, can be computed. This corresponds to a single cell in the design space matrix, the lower left corner of the matrix. Several traces from the Internet traffic archive (LBNL) are anonymized with blackmarking. Please refer to the UCRchive for a comprehensive list of available traces [§].

Truncation ($TR\{t\}$) replaces the t least significant bits of an IP address with 0. Thus, truncating 8 bits would replace an IP address with its corresponding class C network address. With respect to the design space, this means only metrics with a resolution and subset size of [00, 32 - t] can be computed when truncation is used. The number of available granularities decreases with t, the number of truncated bits, as illustrated in Fig. 2. The traces from the Abilene network, which have been used to evaluate numerous anomaly detection approaches, are anonymized with truncation of 11 bits.

Random permutation (RP) translates IP addresses using a random permutation that does not preserve the prefix structure. Since permutation creates a one-to-one mapping, the number of distinct IP addresses is the same. Hence, when random permutation is used for anonymizing a trace, metrics that can be computed on it may only feature the highest and lowest resolution values, as well as largest and smallest subset sizes (see Fig. 2). Note that these correspond to the four corners of the design space matrix. A special case of random permutation is the renumbering of IP addresses (e.g., TCPdpriv with level 0). Packet Traces from UCLA CSD, as well as several traces from the Internet traffic archive (LBNL) are sanitized using random permutation.

Partial prefix-preserving permutation ($PPP\{p\}$), as proposed in [16], permutes the host and network part of IP addresses independently. It preserves the prefix structure in a trace at one specific prefix length p, and at the level of IP addresses. Consequently, this technique retains all granularities that have a resolution and subset size of either 00, p, or 32 (see Fig. 2). PPP is a popular technique that is used for anonymizing traces from the Passive Measurement and Analysis project (PMA) and the Internet traffic archive (LBNL). PMA uses PPP{12} and PPP{16}. Moreover, level 1 of TCPdpriv corresponds to PPP{16}.

§http://networks.cs.ucr.edu/ucrchive/measurement.htm

Prefix-preserving permutation (PP) permutes IP addresses so that two addresses sharing a common real prefix, also share an anonymized prefix of equal length (see e.g., [8]). This is actually the best anonymization technique with respect to utility since it preserves the full design space. We will use it in our measurement study as a reference to a perfect anonymization scheme (with respect to utility). PP is applied to traces from CAIDA and CRAWDAD.

Note that anonymization always involves a tradeoff between data utility and the risk of privacy violations [5]. Ideally, an anonymization scheme would guarantee perfect protection from privacy violations (low risk) without affecting the utility of data with respect to some target application (high utility). In this paper, however, we focus on data utility only. For attacks on anonymization techniques please refer to [4, 10, 3, 17].

3. Methodology

In this section, we describe our methodology for studying the impact of anonymization on statistical anomaly detection. We introduce the data set used in this study, and describe the methodology for classifying it. We further present the Kalman filter that is used as detection algorithm.

3.1. Measurement Data

The data used in this study was captured from the four border routers of the Swiss Academic and Research Network (SWITCH, AS 559) [21], a medium-sized backbone operator, connecting several universities and research labs (e.g., IBM, CERN) to the Internet. The SWITCH IP address range contains about 2.4 million IP addresses and the traffic volume varies between 60 and 140 million NetFlow records per hour. We analyzed a three-week period (from August 19th to September 10th 2007). This data set contains a variety of anomalies with diverse characteristics. In total, 43.2 billion flows covering a volume of 713 Terabytes of traffic were analyzed. In contrast to previous work, this study is based on un-sampled and non-anonymized flow data. Such datasets are difficult to obtain (at least over longer observation periods), but mandatory if bias and distortion in the results are to be avoided.

3.2. Ground Truth

The first step of any measurement study on anomaly detection is the establishment of ground truth for the available traces. Unfortunately, obtaining ground truth for an unclassified data set is still a large challenge and involves a lot of manual inspection. In the following, we describe our methodology for labeling the dataset.

Visual inspection of metric timeseries: We computed the timeseries for five well-known metrics over an entire three weeks period at 15-minute intervals, resulting in 2016 data points per metric. As metrics, we selected byte, packet, and flow counts, unique IP address counts, and the Shannon entropy[¶] of flows per IP address. Moreover, we distinguished incoming and outgoing traffic, as well as TCP and UDP traffic adopting what is common practice in the anomaly detection community. Finally, we visually inspected all these timeseries for unusual events.

Analysis of raw NetFlow traces: For all intervals that could not be classified by timeseries inspection with high confidence, we did further analysis on the raw NetFlow traces. For this purpose,

[¶] $H(X) = -\sum_{i=1}^{n} P(x_i) \cdot log_2(P(x_i))$

	Vol	DoS	Sca	Flu	Unk	Tot
TCP	75	32	539	24	19	689
UDP	64	14	4	239	28	339

Table 2: Ground truth: Number of anomalous intervals per anomaly type and total for UDP/TCP.

we used nfdump [14], a tool developed by SWITCH for forensic analysis to collect more information about suspicious events, e.g., which hosts and ports are affected.

Assigning ground truth to each interval: If at least *one* of the analyzed metric timeseries exposed an unusual event in some interval, we classified that interval as anomalous. Note here that most events were visible across multiple metric timeseries.

Identifying the anomaly type: Having classified all intervals as normal or anomalous, we went one step further and assigned the anomalous events to different types. Since a commonly agreed methodology for classifying known anomalies is not yet established, we define and distinguish the following types of events:

- **Volume**: Volume anomalies are events that cause a sharp increase or decrease in the volume-based metrics, but do not affect the feature-based metrics. In our trace, we found two large loss events and several high-volume flows or alpha flows.

- **(D)DoS**: Denial of Service attacks cause a concentration of the flows on one or few target IP addresses and hence a drop in the destination IP address entropy. If, on top, the attack is distributed, we will additionally see a spike in the source IP address counts and entropy metrics. If the attack is large in terms of flows or even packets, in addition, it will cause a spike in volume-based metrics.

- **Scan**: Scans provoke an increase in the destination IP address counts and entropy. If the attack sources are distributed, we will also see an increase in the source IP counts.

- **Network Fluctuation**: Events that cause an increase or decrease in the IP counts at lower resolutions but are not significant in the IP address counts at the highest resolution, fall into this class. Examples of such anomalies are ingress shifts and route flaps, but also massively distributed coordinated events that involve only a small number of IP addresses (e.g., botnet activity or stealth scans).

- **Unknown**: Despite the classification effort that was made, some events remained unclassified. All unclassified events fall into this class.

Table 2 summarizes the identified events in our three-week long trace for UDP and TCP traffic. Note that we counted the number of anomalous intervals not the number of anomalies. Therewith the number of anomalous intervals can be quite large for anomalies that persist over several hours or even days. For example, a large part of the 542 TCP-scanning intervals belongs to a single long-lasting event. Likewise, most of the 239 intervals classified as *network fluctuation* belong to one single event that reappeared every 2 hours over several days, but lasted each time only for few intervals.

3.3. Anomaly Detection with the Kalman Filter

From the list of available statistical anomaly detection methods, we selected the Kalman filter since its excellent performance for anomaly detection has been shown in [20]. The Kalman filter is an efficient recursive filter that estimates the state of a dynamic system from a series of incomplete

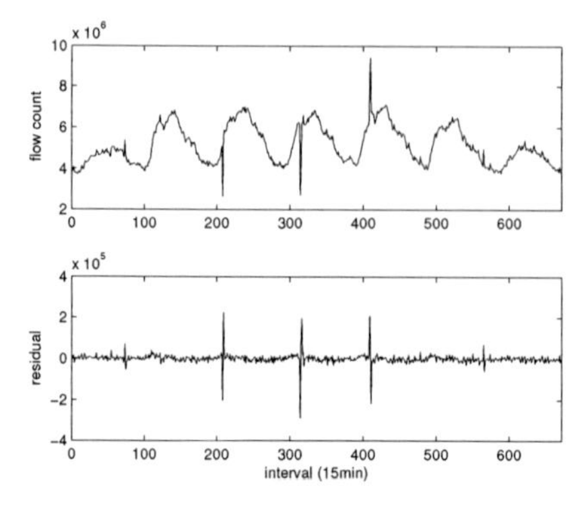

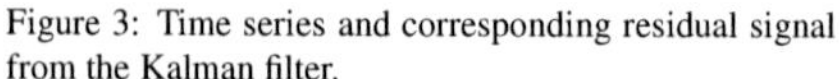

Figure 3: Time series and corresponding residual signal from the Kalman filter.

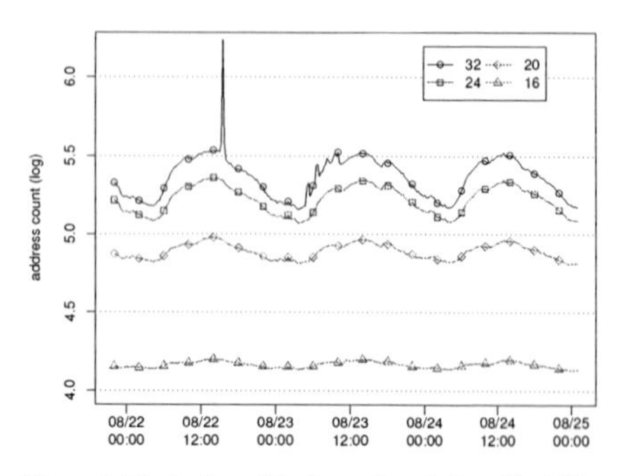

Figure 4: Illustration of the loss of resolution effect. The unique address count at resolutions 32, 24, 20 and 16 is shown.

and noisy measurements. It models normal traffic as a "measurement-corrected" AR(1) process plus zero-mean Gaussian noise. The difference between this model and the actual measured time series, the so-called residual, is used for detection (see Fig. 3 for an illustration). An alarm is raised by the detector if the residual excesses some threshold.

We calculated a total of 60 metrics (see next Section) on the three-weeks data set, and applied the Kalman filter *separately* to each of those. This results in a $\{60 \text{ x } 2016\}$ matrix of residual values, one for each metric and interval. An anomaly is detected if at least one of the 60 residual values for an interval exceeds a threshold. We assess the performance of the Kalman filter with the help of ROC curves [9]. ROC curves plot the rate of false positives against the rate of true positives for a range of thresholds. As thresholds, we use multiples of the a posteriori estimation for the standard deviation (s) of the considered metric. The thresholds range from $0.2s$ (top right corner) to $2.4s$ (bottom left corner). Remind from theory that an interval which exceeds the standard deviation of the noise process is considered unusual. In general, the higher the true positive rate at a particular false positive rate, the better the performance of the detector. Hence, the curve of an optimal detector goes through the top left corner whereas a curve close to the diagonal represents random guessing.

For both UDP (left) and TCP (right) traffic, Fig. 5 shows one ROC curve per anomaly class, and one curve for the overall detection capabilities. We restrict our analysis to a specific type of anomaly as follows: we exclude all intervals that have been manually classified as anomalous, but of a different type, and used these shortened timeseries as input for the detector. We see from this Figure that the Kalman filter generally works well. For UDP, we obtain very high true positive rates at a small false negative rate for all classes of anomalies. Detection results for TCP traffic are slightly worse. This is however expected since TCP has a larger traffic share than UDP and is also more volatile compared to UDP traffic.

When examining the traces in detail, we found that false positives are often due to fast increases or decreases in the normal daily traffic cycle, which are misinterpreted by the Kalman filter as anomalous events. Another source for false positives are temporary increases in the volatility of a metric. This type of false positives can be avoided by recalibrating the Kalman filter each time the volatility changes. We also observed that the Kalman filter tends to miss anomalies that increase or decrease more slowly, and take multiple intervals to grow to their full strength. For some of those, however, the Kalman filter detects the end of the anomaly when the traffic suddenly falls or rises to its previous

	vbm{00}	fbm{16}	fbm{24}	fbm{32}
PP	x	x	x	x
PPP(16)	x	x		x
RP	x			x
TR(08)	x	x	x	
TR(16)	x	x		
TR(32)	x			

Table 3: Metrics available with different anonymization techniques (vbm = volume-based metrics, fbm = feature-based metrics).

level. There is one more thing to point out in Fig. 5: The false positive rate at a specific threshold is practically the same for all classes of anomalies (i.e., all markers representing the same threshold are more or less vertically aligned). The reason therefore is that the false positive rate depends only on the normal traffic, which is the same for all cases, and not on the studied anomaly.

3.4. Computing the Utility of Anonymized Data

Basically, we use the same methodology for the non-anonymized case as for the anonymized traffic. The difference is, however, that we run the Kalman filter only on the subset of the 60 metrics that is available for the anonymization technique under study.

The 60 studied metrics are different variants of three volume-based metrics (vbm) (byte, packet, and flow counts) and two feature-based metrics (fbm) (the unique IP address count, and the Shannon entropy of flows per IP address). We distinguished TCP and UDP traffic as well as incoming and outgoing traffic. Moreover, we used a subset size of 0 for all metrics, i.e., we computed our metrics over all available traffic. Since we were interested in exploring how the restriction of available resolutions affects anomaly detection, we computed the metrics at four representative resolution levels of {00, 16, 24, 32} bits[‖]. Here, we made a distinction between volume- and feature-based metrics. We computed volume-based metrics only at the lowest resolution of all traffic. This is because the computation of volume metrics at higher resolutions (e.g., the volume per IP) results in one time series per entity, and a clustering mechanism would be required to summarize them into one metric. The impact of anonymization on clustering algorithms, however, is not subject of this study. Feature-based metrics, on the other hand, were computed at a resolution of {16, 24, 32} bits. The lowest resolution was not used for feature-based metrics since it results always in a value of one (e.g., there is only one unique /0 prefix in the trace). Therewith, we obtain a total of $(3[vbm] + (2[fbm] \times 2[src/dst] \times 3[res])) \times 2[in/out] \times 2[udp/tcp] = 60$ detection metrics.

The resolutions available with each anonymization scheme are given in Table 3. Also refer to Fig. 2 which illustrates the subset of the design space available with each technique. The volume-based metrics computed at a resolution of 00 bits (vbm{00}) are available for all anonymization techniques. Feature-based metrics computed at a resolution of 16 bits are available with all techniques that retain this resolution, i.e., PP, PPP(16), TR(08), and TR(16). Likewise, feature-based metrics at a resolution of 24 bits are available with all anonymization techniques that retain the resolution of 24 bits, i.e., PP, and TR(08). Finally, feature-based metrics computed at a resolution of 32 bits are available with all permutation-based techniques since these retain the notion of individual IP addresses.

To assess the utility, we compared the ROC curves obtained when using the restricted set of metrics available with each anonymization technique. Further, we reduced the complex ROC curves

[‖] A resolution of 8 bits is too low for our data set.

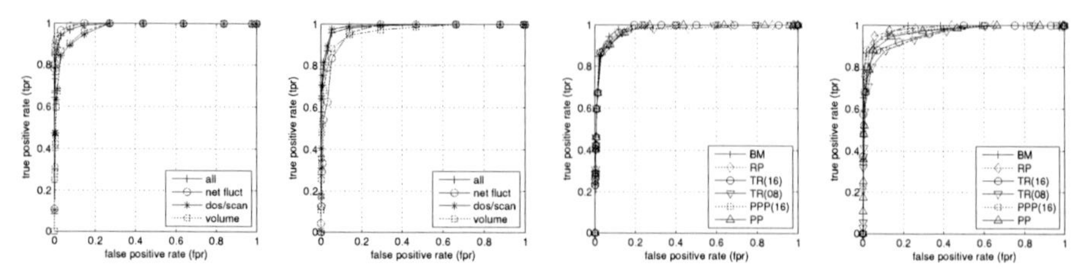

Figure 5: ROC curves for different types of anomalies in UDP (left) and TCP traffic (right).

Figure 6: Volume anomalies in *anonymized* traffic, UDP (left), TCP (right).

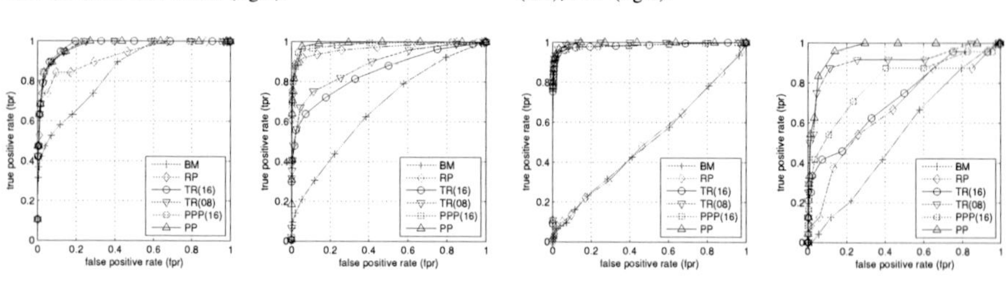

Figure 7: Scanning and denial of service anomalies in *anonymized* traffic, UDP (left), TCP (right).

Figure 8: Network fluctuations in *anonymized* traffic, UDP (left), TCP (right).

to a single utility value by computing the area under the curve (AUC) [1]. To obtain the AUC from the empirical ROC curves, we fitted a piecewise cubic Hermite interpolating polynomial to the data, and approximated the area under the curve numerically. In the next section, we describe and discuss the results obtained with the methodology described above.

4. Measurement Results

We commence this section with an illustrative example for the *loss of resolution* effect, i.e., we examine how the restriction of available resolutions through anonymization impacts the detectability of anomalies. In Fig. 4, we plot the count of unique source addresses in all incoming TCP traffic at different resolutions of IP addresses (32), /24 networks (24), /20 networks (20), and /16 networks (16). In the curve for the highest resolution value, we see a large peak corresponding to 1.2 million additional IP addresses launching a denial of service attack. Interestingly, the peak completely disappears at a resolution of 24 and lower. Observing the curve for /24 networks more closely, we find a very small, non-significant peak of $\approx$ 5'000 additional networks in the same interval. From this observation, we conclude that the attack sources remain in a few /24 networks. As seen in the figure, this example anomaly disappears at lower resolutions, and thus, it will be hard or impossible to detect in data anonymized with truncation. In the following, we systematically assess the overall utility of the different anonymization techniques with the help of the Kalman filter detector applied to the whole three-week long data set.

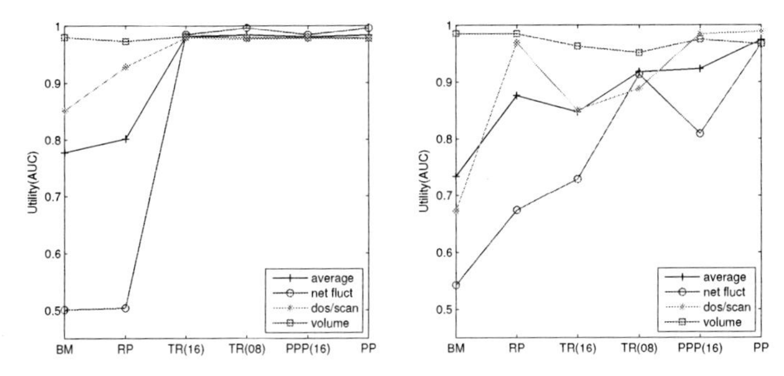

Figure 9: Utility for large-scale anomaly detection for different anonymization techniques. Left plot is for UDP traffic, right plot is for TCP traffic.

4.1. ROC Curves for Anonymized Data

To assess the utility of the different anonymization schemes, we study the impact of anonymization separately for each type of anomaly. This is necessary since each type of anomaly is exposed at characteristic traffic granularities, and thus, is impacted differently by the restriction of the granularity design space through anonymization. To give an example: Alpha flows are mainly visible in byte and packet counts computed at the lowest resolution, while scans are primarily visible in feature-based metrics at higher resolutions. Hence, studying all anomalies together will thus not lead to any conclusive results. As a result, we distinguish the following anomalies for our evaluation:

Volume Anomalies, such as outage events and alpha flows, are mainly exposed by volume-based metrics. Since volume-based metrics at the lowest resolution are available with all anonymization schemes (see Table 3), we expect that anonymization does not have a large impact on the detection of volume anomalies. Indeed, the measurement results presented in Fig. 6 clearly confirm this expectation. Anonymization does not alter the utility of data when one is solely interested in detection of volume anomalies. When examining the plot for TCP traffic more closely, we observe that blackmarking and random permutation perform slightly better than the other schemes. We conclude that using fewer metrics might even be beneficial as it results in fewer false positives.

Scanning and denial of service anomalies are both mainly visible in feature-based metrics. Measurement results for this class of anomalies are presented in Fig. 7. The curves for UDP and TCP confirm that blackmarking performs worst, whereas prefix-preserving permutation (which does not restrict the granularity design space at all) has the best performance. To give an example: at a false positive rate of 0.02, the detection rate for blackmarking is reduced by 50% for UDP, and even more for TCP traffic. Surprisingly, the ranking for truncation and random permutation is not the same for UDP and TCP traffic. For TCP traffic, random permutation outperforms truncation, while for UDP traffic the opposite applies. We think this is due to structural differences between, normal and anomalous, UDP and TCP traffic. We verified on the data that TCP scans and DoS attacks are mainly visible at the resolution of individual IP addresses, which are preserved by random permutation but not truncation. On the contrary, UDP scans and DoS attacks are visible at high and low resolutions, but metrics at lower resolutions have fewer false positives. Consequently, truncation outperforms random permutation for detecting UDP scans and DoS attacks.

Network fluctuations are mainly visible in feature-based metrics at lower resolutions. The ROC curves for UDP and TCP traffic presented in Fig. 8 show that detection of network fluctuations is almost impossible when either blackmarking or random permutation is used. Truncation of 8 bits, on the other hand, does not result in a severe performance degradation, neither for TCP nor UDP traffic. UDP and TCP differentiate with respect to the performance of 16-bit truncation and PPP. This might, however, be a particularity of our data set: Most of the 239 network fluctuation anomalies in the UDP traffic are mainly visible at a resolution of 16 bits, whereas TCP network fluctuations are rather visible at a resolution of 8 bits.

4.2. Utility of Anonymized Traces for Anomaly Detection

We will now summarize the detailed results from Section 4.1 using the area under the curve (AUC) as a measure for the utility of an anonymized data set. An AUC value of 1 means that the detector achieves perfect accuracy, whereas a detector with an AUC of 0.5 is not useful. In Fig. 9, we plot for each anonymization technique (x-axis) the AUC value (y-axis). The left plot summarizes the results for UDP traffic and the right plot is for TCP traffic. We show one curve for each of the three anomaly classes (volume, scan/DoS, and network fluctuations), as well as the average over all classes. Note that this average utility corresponds to a data set that contains the same ratio of volume, scan/DoS, and network fluctuation anomalies.

Fig. 9 clearly shows that the detectability of volume anomalies is not impacted by anonymization. Furthermore, it confirms our intuition that prefix-preserving anonymization has the best utility with respect to backbone anomaly detection in general, and that blackmarking has the worst utility for all classes except volume anomalies. The utility of random permutation, truncation, and partial prefix-preserving permutation largely depends on the type of anomaly in question. Partial prefix-preserving permutation performs almost as well as prefix-preserving permutation; it has a lower utility only for network fluctuation anomalies in TCP traffic. Random permutation has a high utility for the detection of large-scale scans and denial of service attacks, but a low utility for detecting network fluctuations in UDP as well as TCP traffic. Truncation performs very well for UDP traffic in our measurements, but has a lower utility for detecting scans, denial of service attacks, and network fluctuation anomalies in TCP traffic. Moreover, the utility for these anomalies decreases when more bits are truncated.

Naturally, the overall utility of an anonymization technique applied to a given data set depends on the mix of anomalies within the trace. For our data set, the overall utility for detecting TCP anomalies is clearly dominated by the 542 scanning intervals. Likewise, the overall utility for UDP traffic is to a large extent dominated by the 239 intervals with network fluctuations. Hence, for this anomaly mix, PP and PPP offer the highest utility, truncation lies in the middle, and random permutation and blackmarking result in the lowest utility. We summarize the results for our data set as follows:

- Blackmarking decreases the utility for detecting anomalies in UDP and TCP traffic dramatically.
- Random permutation performs very bad with the detection of anomalies in UDP traffic, while preserving the utility for TCP traffic.
- Truncation of 8 or 16 bit decreases the utility for detecting anomalies in TCP traffic by roughly 10 percent, while performing well for UDP traffic.

- (Partial) prefix-preserving permutation has no significant negative impact on the utility for detecting anomalies in UDP and TCP traffic.

To derive more general conclusions about the utility of different anonymization schemes it would be helpful to study the anomaly mix in available flow traces. If this mix converges, at least for traces recorded in the same time period, a general conclusion about the utility can be derived directly from our results. Moreover, we expect that the results for volume, scan/DoS, and network fluctuation anomalies hold also for other traces. Verifying this assumption requires that we apply our methodology to further un-anonymized data sets from different networks. Unfortunately, such traces are not currently available in abundance.

5. Implicit Traffic Aggregation

In Section 4, we investigated how anomaly detection results are falsified when valuable resolutions are not available for anomaly detection. Another aspect of anonymization that is worth studying is the restriction along the subset size dimension, causing an implicit aggregation of traffic.

Let us illustrate this with an example. Consider the case where traffic from a single host (i.e., subset size 32) is to be investigated in presence of 4 bits truncation. The best one can do under these circumstances is investigation of the /28 network that contains the host, since the individual host can no longer be distinguished from other hosts in the same network. As a consequence, the analyzed traffic is a mixture of traffic from the target host and traffic from other hosts in neighboring subnets. In accordance with loss of resolution, we refer to this effect as the *loss of focus effect.*

The impact of the loss of focus effect is, of course, highly dependent on the distribution of traffic in the studied network. *It is extremely difficult to predict the implications for a particular case.* In the worst case, even truncation of a single bit can be fatal. That is, traffic characteristics of the target host could get lost completely in another host's traffic. In the best case, where the truncated subnet is dedicated to a single host, no traffic is aggregated. In that case, the loss of focus effect is simply reduced to a loss of resolution effect.

We can, however, estimate the average severity of truncation-induced traffic aggregation by analyzing the count of additional, non-belonging, flows for 170 individual hosts. In particular, we count the flows of 170 webservers belonging to a single /16 network in our un-anonymized traces. Then, we apply truncation to the traces and count the number of flows to the subnets containing the webservers.

When truncating a single bit, more than 50% of the observed webservers experience no additional traffic. Only around 10% of the webservers have a resulting traffic increase of 100% or more. However, if more bits are truncated the situation gets worse. The ratio of unaffected servers drops to 20% for 2 bits, 5% for 4 bits, and even 0% for 8 bits. Similarly, the ratio of servers that experience at least a doubling of traffic goes up to 25% for 2 bits, 55% for 4 bits and 89% for 8 bits.

We conclude that accurate detection of small-scale anomalies** is very difficult, if not impossible, when the desired subset size is not supported by the restricted granularity design space. The probability that host characteristics are lost in aggregated traffic is simply too high. Only anomalies of sufficient scale have a chance to be visible in aggregated traffic at larger subset sizes. In addition, false positives are introduced by the aggregation with traffic from other hosts.

**With small-scale anomalies we denote anomalies affecting only a single host or a small subnet.

6. Conclusion

In this paper, we have answered the question of how anonymization techniques impact statistical anomaly detection. We introduced the detection granularity design space as an important tool to illustrate this impact. We have shown how the design space, spanned by subset size and resolution, is reduced by the most common IP address anonymization techniques. Finally, we analyzed the utility of anonymized traces for the detection of large-scale anomalies, as well as the impact on traffic characteristics of individual hosts with the aid of backbone traffic traces gathered over a three-weeks period.

In general, our results indicate that the restriction of the granularity design space through anonymization hinders anomaly detection. With respect to the individual techniques, we have found that prefix-preserving permutation offers the best utility, and blackmarking performs the worst. Moreover, we have shown that the performance of random permutation, partial-prefix preserving permutation, and truncation strongly depends on the type of anomaly that is studied, as well as the underlying transport protocol. Our results indicate that the detection of volume anomalies, such as outages or alpha flows, is not impacted by anonymization at all. The utility for detecting scans and denial of service attacks degrades when truncation is applied. Detection of network fluctuations, on the other hand, is impacted principally by blackmarking and random permutation.

Thus, if one is interested in a particular type of anomaly, anonymization could be tuned in a way such that the results are less impaired. In addition, we have shown that the anonymization-induced loss of focus, i.e., when the desired subset size is not available, in most cases completely distorts the traffic characteristics of individual hosts.

While we provided some interesting insights on the impact of anonymization techniques on anomaly detection, we encourage further research with un-anonymized traffic traces to challenge or confirm our results.

Acknowledgments

We are grateful to Elisa Boschi from Hitachi Europe for the numerous valuable discussions and to SWITCH for providing the traffic traces used in this study.

REFERENCES

[1] BRADLEY, A. The use of the area under the ROC curve in the evaluation of machine learning algorithms. *Pattern Recognition 30* (1997), 1145–1159.

[2] BRAUCKHOFF, D., TELLENBACH, B., MAY, M., AND LAKHINA, A. Impact of packet sampling on anomaly detection metrics. In *ACM SIGCOMM Internet Measurement Conference (IMC)* (25-27 October 2006), pp. 159–164.

[3] BREKNE, T., ÅRNES, A., AND ØSLEBØ, A. Anonymization of IP traffic data: Attacks on two prefix-preserving anonymization schemes and some proposed remedies. In *Workshop on Privacy Enhancing Technologies* (2005), pp. 179–196.

[4] COULL, S., WRIGHT, C., MONROSE, F., COLLINS, M., AND M.K.REITER. Playing devil's advocate: Inferring sensitive information from anonymized network traces. In *14th Annual Network and Distributed System Security Symposium* (February 2007).

[5] DUNCAN, G. T., KELLER-MCNULTY, S. A., AND STOKES, S. L. Disclosure risk vs. data utility: The r-u confidentiality map. Tech. Rep. 121, National Institute of Statistical Sciences, December 2001.

[6] EU. Directive 95/46/ec of the European parliament and of the council. OJ L 281, 23.11.1995, p. 31, October 1995.

[7] EU. Directive 2002/58/ec of the European parliament and of the council. OJ L 201, 31.07.2002, p. 37, July 2002.

[8] FAN, J., XU, J., AMMAR, M. H., AND MOON, S. B. Prefix-preserving IP address anonymization. *Comput. Networks 46*, 2 (2004), 253–272.

[9] FAWCETT, T. An introduction to ROC analysis. *Pattern Recognition Letters 27*, 8 (2006), 861–874.

[10] KOUKIS, D., ANTONATOS, S., AND ANAGNOSTAKIS, K. G. On the privacy risks of publishing anonymized IP network traces. In *Communications and Multimedia Security* (2006), vol. 4237 of *Lecture Notes in Computer Science*, Springer, pp. 22–32.

[11] LAKHINA, A., CROVELLA, M., AND DIOT, C. Diagnosing Network-Wide Traffic Anomalies. In *ACM SIGCOMM* (Portland, August 2004).

[12] MAI, J., SRIDHARAN, A., CHUAH, C.-N., ZANG, H., AND YE, T. Impact of packet sampling on portscan detection. *Selected Areas in Communications, IEEE Journal on 24*, 12 (Dec. 2006), 2285–2298.

[13] MINSHALL, G. Tcpdpriv. http://ita.ee.lbl.gov/html/contrib/tcpdpriv.html.

[14] NFDUMP. http://nfdump.sourceforge.net/.

[15] OHM, P., SICKER, D., AND GRUNWALD, D. Legal issues surrounding monitoring during network research (invited paper). In *ACM SIGCOMM Internet Measurement Conference (IMC)* (2007).

[16] PANG, R., ALLMAN, M., PAXSON, V., AND LEE, J. The devil and packet trace anonymization. *SIGCOMM Comput. Commun. Rev. 36*, 1 (2006), 29–38.

[17] RIBEIRO, B., CHEN, W., MIKLAU, G., AND TOWSLEY, D. Analyzing privacy in enterprise packet trace anonymization. In *15th Annual Network & Distributed System Security Symposium (NDSS 08)* (February 2008).

[18] SLAGELL, A., LAKKARAJU, K., AND LUO, K. Flaim: A multi-level anonymization framework for computer and network logs. In *20th USENIX Large Installation System Administration Conference (LISA'06)* (2006).

[19] SOULE, A., LARSEN, H., SILVEIRA, F., REXFORD, J., AND DIOT, C. Detectability of traffic anomalies in two adjacent networks. In *Passive And Active Measurement Conference (PAM)* (2007).

[20] SOULE, A., SALAMATIAN, K., AND TAFT, N. Combining filtering and statistical methods for anomaly detection. In *IMC '05* (2005).

[21] SWITCH. The swiss education and research network. http://www.switch.ch.

[22] WEI, S., MIRKOVIC, J., AND KISSEL, E. Profiling and clustering internet hosts. In *2006 International Conference on Data Mining* (2006).

[23] XU, K., ZHANG, Z.-L., AND BHATTACHARYYA, S. Profiling internet backbone traffic: behavior models and applications. In *SIGCOMM 2005* (2005).

[24] YURCIK, W., WOOLAM, C., HELLINGS, G., KHAN, L., AND THURAISINGHAM, B. Privacy/Analysis Tradeoffs in Sharing Anonymized Packet Traces: Single-Field Case. In *Third International Conference on Availability, Reliability and Security (ARES)* (2008).

[25] ZHANG, J., BORISOV, N., AND YURCIK, W. Outsourcing security analysis with anonymized logs. In *Securecomm and Workshops, 2006* (Aug. 28 2006-Sept. 1 2006), pp. 1–9.

Key words – DiffServ, UTRAN, Iub, dimensioning, IP transport

Xi Li[1], Wojciech Bigos[2], Carmelita Goerg[1], Andreas Timm-Giel[1] and Andreas Klug[3]

DIMENSIONING OF THE IP-BASED UMTS RADIO ACCESS NETWORK WITH DIFFSERV QOS SUPPORT

Dimensioning is an essential task of network engineering to provide multi-service QoS guarantees without overprovision, i.e. to achieve efficient bandwidth utilization while assuring the QoS requirements of different services. This paper presents analytical models for dimensioning transport bandwidth on the Iub interface in the IP-based UMTS Terrestrial Radio Access Network (UTRAN), where the deployed QoS architecture is based on Differentiated Service (DiffServ) using an integrated Weighted Fair Queue (WFQ) and Strict Priority (SP) scheduling. The analytical models for the Iub bandwidth dimensioning are validated by the simulations. The simulation results demonstrate that the proposed analytical methods can appropriately estimate the application performances of the different classes based on DiffServ QoS framework, hence we can efficiently plan bandwidth capacity of the network links and promise multi-service QoS provisioning in the UTRAN.

1. INTRODUCTION

Third generation (3G) mobile communication systems, in particular the Universal Mobile Telecommunication Systems (UMTS), are expected to have an intensive growth in the next few years caused by a continuously increasing number of mobile subscribers and operative networks all over the world, as well as by a dramatically growing traffic demand for data applications like video streaming, web and multimedia services. This in turn requires the Universal Terrestrial Radio Access Network (UTRAN) to offer much higher transport capacity for supporting the evolved UMTS radio interface and newly introduced HSDPA (High Speed Downlink Packet Access) and HSUPA (High Speed Uplink Packet Access) services [1][2]. But in the most current deployed ATM based UTRAN transport network, increasing the ATM capacity by leasing additional E1/T1 lines leads to a linear increase of the operation expense. Therefore, there is a strong trend of using IP and Ethernet as underlying transport technologies replacing the current ATM transport to bring cost savings in transport.

IP-based UTRAN is introduced and studied in 3GPP Release 5 [3] and MWIF [4]. The main benefits of pursuing IP-based transport are: (1) Low cost IP equipment and cheap IP bandwidth

[1] Communication Networks, University of Bremen, Germany
email: [xili | cg| atg]@comnets.uni-bremen.de
[2] Nokia Siemens Networks Sp. z o.o., Wroclaw, Poland
email: wojciech.bigos@nsn.com
[3] Nokia Siemens Networks GmbH & Co. KG, München, Germany
email: andreas.klug@nsn.com

based on the Ethernet transport significantly reduces the infrastructure costs; (2) IP is widely deployed, has high flexibility and scalability; (3) IP facilitates the integration of different radio access technologies operating over a common IP backbone and therefore enables the development of heterogeneous network access and also can better cope with IP-based 3G core networks. The employment of IP in the transport network of UMTS is an essential step towards one "All-IP" network merging the fix and the mobile networks.

Despite prominent technique advantages and low costs of deploying IP, the most important challenge of an IP-based UTRAN network is how to provide and differentiate the QoS for a wide range of diverse services with various QoS requirements. The QoS challenge is mainly associated with the fact that IP is designed for the `best effort' Internet where there is no guarantee for the QoS. To provide superior QoS support in the IP-based UTRAN, an efficient IP QoS mechanism should be applied to distinguish different service flows and provide the desired level of service. The Internet Engineering Task Force (IETF) has developed a number of IP QoS schemes, most notably, Differentiated Services (DiffServ) [5]. It is considered as a promising QoS support mechanism to be deployed in the IP-based UMTS network.

In this paper, we use a sophisticated QoS structure in the IP-based UTRAN, which is based on DiffServ QoS scheme with an integrated Weighted Fair Queue (WFQ) and Strict Priority (SP) scheduling. And based on this QoS structure, analytical dimensioning models are proposed for the dimensioning of the Iub interface, i.e. the interface between Node B and RNC (Radio Network Controller) within UTRAN, for supporting a wide range of services each with specific QoS requirement. As the transport resources of the radio access network are quite limited and expensive, effective bandwidth utilization in the UTRAN, especially on the Iub interface is directly related to the transport costs. Thereby, the goal of the Iub dimensioning is to guarantee the required QoS of various services in a cost efficient way in terms of achieving a high transmission efficiency and maximal utilization of the Iub bandwidth.

The remainder of the paper is organized as follows: Section 2 gives an overview of related work. Section 3 describes the DiffServ QoS scheme. Section 4 introduces the DiffServ based QoS architecture deployed in the IP-based UTRAN in this paper. Section 5 presents the proposed analytical dimensioning methods. Section 6 validates the proposed analytical models by simulations. Section 7 analyzes the impact of DiffServ QoS support on the dimensioning of the transport network resources with simulations. The end gives the conclusions.

2. RELATED WORK

The IETF has developed a number of different QoS mechanisms for IP networks to support multi-service, like Integrated Services (IntServ) [6] with Resource Reservation Protocol (RSVP) [7], DiffServ, Multiprotocol Label Switching (MPLS) [8], etc. But they are designed for fixed networks (e.g. Internet) and work inefficiently in mobile environments. Thus, discussions are raised on how to integrate them into a coherent fully IP-based mobile network architecture. [9] studied various IP QoS mechanisms and proposed an architecture for provisioning of QoS in IP-based mobile access networks. [10] presented a framework and basic methodology for deploying DiffServ in UMTS/GPRS backbone networks to support the core network's QoS requirements. However, the issue of network dimensioning has not been addressed too much when such IP QoS schemes and architectures are applied in the IP-based mobile networks. To the best knowledge of the authors, there were not many studies on the dimensioning of the IP-based mobile access networks which deploy IP QoS architecture like DiffServ. This paper intends to discuss a DiffServ-based UMTS access network, and propose analytical dimensioning methods for a cost-efficient allocation of the expensive transport resources at the UTRAN Iub interface, to efficiently transmit elastic traffic and streaming traffic types.

3. INTRODUCTION OF DIFFSERV

DiffServ is an IP-based QoS support framework. It provides different treatment to flows and aggregates flows by mapping multiple flows into a finite set of service levels, namely Per Hop Behavior (PHB) groups. Each PHB is identified by a DiffServ Code Point (DSCP) in the IP header which provides information about the QoS requested for a packet, and defines a specific forwarding treatment that the packet will receive at each node. DSCP are defined in IETF RFC 2474 [11]. It is specified in the 8-bit Type of Service (TOS) field in IPv4 header (in Traffic class Header field in IPv6 header). The DSCP enables network routers to handle IP packets differently depending on the code point and hence their relative priority.

3.1. DIFFSERV PHBS

The DiffServ working Group of IETF has defined a number of different PHB groups for different applications. There are three most common PHBs defined in the IETF, Best Effort (BE) PHB, Expedited Forwarding (EF) PHB and Assured Forwarding (AF) PHB.

BE PHB is for the traditional Internet traffic and its usage implies that the nodes in the path will do their best to forward the packet in a fair manner, however, there is no guarantee on its delivery or its level of service. Everybody gets the service that the network is able to provide. Its recommended code point (DSCP) is 000000.

EF PHB is aimed for a low loss, low latency, low jitter, assured bandwidth edge-to-edge service through IP Diffserv domains. It can be understood as a virtual leased line service. Therefore, the bandwidth can not be exceeded but the user can leave it idle or use it to the full extent of its capacity. The holder of this pipe should not be affected by the presence or absence of other users. An example of EF service is voice telephony. The DSCP of EF PHB is 101110, and therefore, there can be only one instant of EF PHB in the IP DiffServ domain.

AF PHB does not provide a bandwidth guarantee but packets are given a higher priority to be transmitted over the network than the packets from the BE PHB. In congestion situations the user of the Assured Forwarding service should encounter less bandwidth decrease than BE PHB users. AF PHB group provides four independently forwarded AF classes. Each of these classes has three levels of dropping precedence: low, medium, and high. Therefore, 12 instances of AF with recommended DSCPs can exist in a DiffServ domain. Their defined DSCP codes are given following:

Low drop: 001010 010010 011010 100010
Medium drop: 001100 010100 011100 100100
High drop: 001110 010110 011110 100110

In each Diffserv node, each AF class is allocated a certain amount of forwarding resources (buffer space and bandwidth).

3.2. DIFFSERV ROUTER

A DiffServ router consists of five components shown in Fig.1. On arrival, a packet is firstly classified by the classifier according to the bilateral service level agreement (SLA). Afterwards, the classifier forwards the packet to the traffic conditioner. The traffic conditioner may include a meter, a marker, a shaper, and a dropper. If accepted, the packet is enqueued into a corresponding buffer and then transmitted according to a specific scheduler policy.

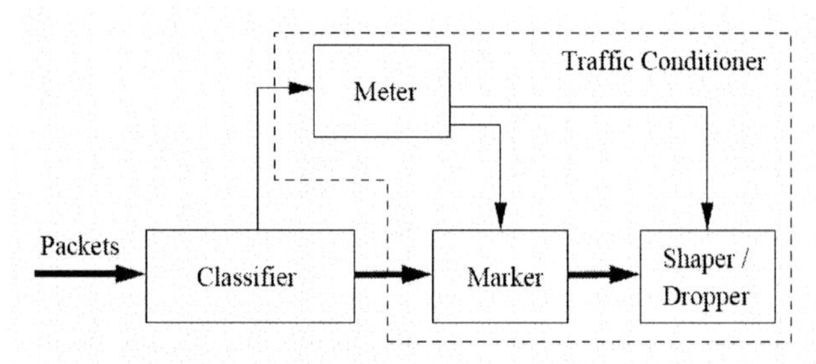

Fig. 1. IP DiffServ Router Building Blocks [9]

Mapping Function: Mapping is the function of translating the QoS of one system to the QoS parameters understood by another system. It is a necessary function for a UMTS packet entering the DiffServ-aware network, or when a packet coming from an external network enters the UMTS network. This function is usually integrated with the classifier into one functioning block. In the IP-based Iub domain (at the Edge Router), it is responsible for translating the QoS of UMTS system to the corresponding IP DiffServ PHB, and vice versa.

Classifier: the packets are classified according to DSCP in the IP header. The classifier has a single input and splits the incoming stream into a number of outgoing ports in terms of different PHB aggregates.

Meters measure the temporal properties of the stream of packets selected by a classifier against a traffic profile. A meter passes state information to other conditioning functions to trigger a particular action for each packet, which is either in-profile or out-of-profile.

Markers set the packet with a particular code point, adding the marked packet to a particular DiffServ behavior aggregate. An existing marking may be changed. It may also use the statistics gathered by a meter to decide on the level of a packet's conformance to its allocated rate characteristics. For a non conformant stream, it takes action by re-marking the packet in either a class with lower priority, or lower drop precedence.

Shapers delay some or all of the packets in a traffic stream in order to shape the stream into compliance with a traffic profile.

Droppers discard some or all of the packets in a traffic stream in order to bring the stream into compliance with a traffic profile. There are two major kinds of dropping elements in a DiffServ router: absolute and algorithmic.

- An absolute dropper is simply used to drop the overloaded traffic of a real-time application, e.g. for the EF PHB, because an overly delayed packet is no longer useful, and is better discarded. Therefore, when congestion happens, the real-time packets are forwarded to this block in the DiffServ node. When real-time traffic misbehaves, the overloaded traffic is sent to the absolute dropper.
- In contrast, an algorithmic dropper drops a packet according to a specific algorithm, usually well before the queue is full. Random Early Detection (RED) [12] is such an algorithmic dropper, suggested for avoiding congestion in the network. It detects the incipient congestion and notifies the congestion manager of the source to decrease its flow rate. This way, congestion is avoided. To perform RED on traffic classes, Weighted RED (WRED) is used to

drop packets selectively based on different priorities specified by IP precedence. Packets with a higher IP precedence are less likely to be dropped than packets with a lower precedence. Thus, higher priority traffic is delivered with a higher probability than lower priority traffic.

Scheduler: The scheduling has the most impact on the level of service a packet receives. It decides on which queue, among a set of queues, to service next, for example Weighted Fair Queuing (WFQ), strict priority queuing.

Queuing: Besides, there is also queuing function in the DiffServ router. They store the packets while they are waiting to receive service by the scheduler and depart to the next hop or DiffServ module.

3.3. WEIGHTED FAIR QUEUING

Weighted Fair Queuing (WFQ) [13] is a popular scheduling algorithm to guarantee bounded delay, guaranteed throughput, and fairness. It is commonly used in a DiffServ node to provide different priorities for different traffic classes. The WFQ scheduling discipline offers QoS by adding a weight to the queues to assign different priority levels for different queues. Each queue is assigned a weight that determines the share of the available capacity that it will have and thus all queues share the bandwidth proportional to their weights. Traffic may be prioritized according to the packet information in the source and destination IP address fields, port numbers and information in the ToS field (e.g. DSCP). The WFQ discipline weights the traffic so that the low-bandwidth traffic gets a fair level of priority. This prevents high-bandwidth traffic from grabbing an unfair share of resources.

Let w_k be the weight of the kth queue and BW the total available IP bandwidth. If there are in total N queues and all queues are transmitting data, then the kth queue obtains a fraction of the total capacity BW_k as calculated in equation (1). But if one priority queue is empty (i.e. not utilizing its allocated bandwidth), then its spare bandwidth shall be fairly shared among the other queues according to their weights.

$$BW_k = \frac{w_k}{\sum_{i=1}^{N} w_i} \cdot BW \qquad (1)$$

4. DIFFSERV QOS STRUCTURE IN IP-BASED UTRAN

This section describes the QoS structure that is studied in this paper for the IP-based UTRAN based on DiffServ QoS scheme.

4.1. MAPPING OF UMTS QOS TO DIFFSERV PHBS

UMTS defines four QoS classes: conversational, streaming, interactive and background class. Both the conversational and streaming class require a certain reservation of resources in the network, mainly intended for carrying Real Time (RT) traffic flows, such as video telephony and audio/video streams. The Interactive and Background class are so called Non Real Time (NRT) traffic. They are mainly meant for carrying elastic Internet applications like web, Email, and FTP.

In this paper, our mapping of the above four UMTS QoS classes to IP DiffServ PHBs is illustrated in Fig. 2. As shown, both the conversational class and streaming class are mapped to EF

PHBs as they require low loss, low latency, low jitter, assured bandwidth, end-to-end service through IP transport domains. The interactive class is mapped to AF PHBs e.g. web traffic. In total, we can set four AF Service classes and each of these classes can also has three levels of dropping precedence: low, medium, and high. We can set different AF PHBs for different UMTS Radio Access Bearer (RAB) types which provide various QoS for the end users. The background class is mapped to BE PHB as it belongs to Best Effort traffic, which has low requirement on the QoS and utilizes the resources that the network is able to provide. For example, best effort HSPA traffic can be set to BE PHB. It should be noted that there can be other alternative mappings of UMTS QoS class to DiffServ PHBs, depending on the system requirements and traffic profiles.

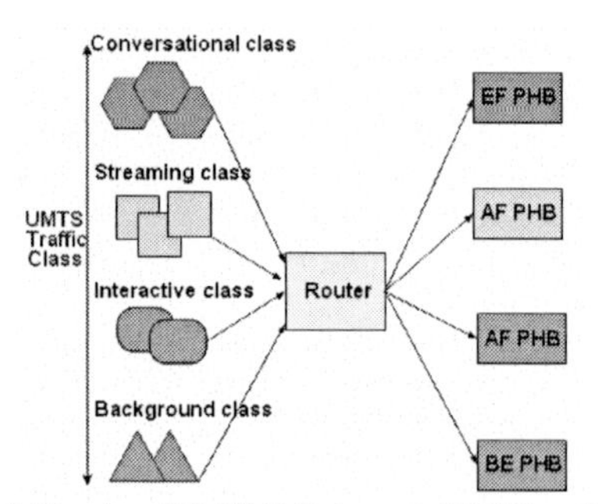

Fig. 2. Mapping of UMTS QoS classes to IP DiffServ PHBs

4.2. QOS STRUCTURE BASED ON DIFFSERV

Based on DiffServ QoS scheme, a refined QoS structure is used in this paper to deploy in the IP-based UTRAN. This structure is shown in Fig.3.

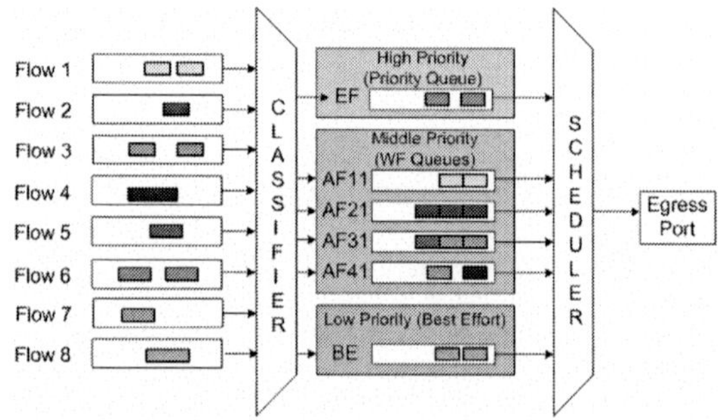

Fig. 3. QoS structure in the IP-based UTRAN – combining priority queue with WFQ

In this proposed structure, there are totally six IP DiffServ PHBs: 1 EF PHB for sending signaling, conversational voice and video streaming traffic; 4 AF PHBs for Interactive traffic class with 4 different priorities (AF11, AF21, AF31 and AF41); 1 BE PHB for background traffic class. Each DiffServ PHB has its own dedicated queue. According to the priorities of different PHBs, a combination of Strict Priority scheduler (SP) and Weighted Fair Queuing scheduler (WFQ) is used in this structure. The EF PHB is given the highest priority, therefore between EF and other PHBs is the Priority queuing. While all AF PHBs and BE PHBs use the spare bandwidth which is not reserved by the EF PHB. The Weighted Fair (WF) Queus are used for AF and BE PHBs and for different AF PHBs and BE PHBs different weights are assigned which determine their priorities, and WFQ scheduling is used to allocate the bandwidth according to their weights among all AF and BE PHBs.

WRED is applied for dropping packets for AF and BE PHBs. As introduced earlier, it is the probabilistic discard of packets as a function of queue fill before overflow conditions are reached. Random dropping takes place when the algorithm detects signs of congestion. The congestion variable is the average queue size that is calculated from the instantaneous queue size by using exponential averaging. For each AF or BE PHB queue, a set of parameters are defined which determine the packet discard probability:

1) max_p: maximum dropping probability;
2) min_{th}: minimum average queue length threshold for packet dropping;
3) max_{th}: maximum average queue length threshold;
4) exponential weight factor (exp.w.) to calculate average (smoothened) queue length.

EF PHB applies the CAR (Committed Access Rate) rate limiting function to drop the overloaded traffic of real-time applications. In this work, we consider only single Iub scenario. The assigned IP bandwidth on this Iub interface determines the shaping rate.

5. ANALYTICAL DIMENSIONING APPROACHES

The applications are classified as the essential categories of streaming flows and elastic Internet traffic. Accordingly, we propose different analytical dimensioning models for transferring these two traffic types within the framework of the IP-based Iub using the above described DiffServ-based QoS structure.

5.1. DIMENSIONING FOR STREAMING TRAFFIC

Streaming traffic in UMTS is usually circuit-switched type of traffic, e.g. voice telephony. The relevant QoS criterion for the streaming services is the blocking probability as a result of call admission control (CAC). Each streaming connection reserves a certain bandwidth for guarantying its data transmission. CAC is used to decide whether there are sufficient free resources on the link to allow a new connection. A connection can only be accepted if adequate resources are available to establish the connection with promised QoS while the agreed quality of service of the existing connections in the network must not be affected by the new connection, i.e. the free bandwidth left by the link shall be larger than the requested bandwidth of the new connection. Therefore the link can only offer a maximum number of simultaneous steaming connections (trunks) each occupies certain bandwidth which is reserved for that connection. The general dimensioning model for RT Streaming service is Multi-Dimensioning-Erlang (MD-Erlang) model, assuming the arrival of the streaming traffic following Poisson process. When there is only one RT streaming traffic type in the network, then the MD-Erlang model is simplified to Erlang-B model. The MD-Erlang model has been studied for the traditional circuit-switched telecommunication network and it is proven to be an accurate mathematic model to derive the necessary link capacity to carry the streaming traffic based on the offered traffic (Erlang) and the occupied bandwidth of a single user at flow level. So in our work, the required IP bandwidth for transmitting the streaming traffic in the IP-based UTRAN is computed by MD-Erlang formula under consideration of a given blocking probability.

5.2. DIMENSIONING FOR ELASTIC TRAFFIC

The elastic traffic is carried by the TCP protocol, where the rate of TCP flow adjusts itself to fill the available bandwidth according to the network traffic condition by using the TCP flow control. The file transmission rates are controlled by the TCP feedback algorithm as a network congestion control function. If TCP works ideally (i.e. instantaneous feedback), all elastic traffic flows going over the same link will share the bandwidth resources equally, and thus the system only carrying elastic traffic flows is essentially behaving as a Processor Sharing (PS) queue, so that the great files will not delay the small ones too much and so that there will be some fairness between the files on each link .

For the elastic traffic (i.e. AF&BE PHBs), the application delay (or throughput) is the relevant QoS criterion. Thus, the task of dimensioning is to determine the minimum required bandwidth, under which a predefined application delay requirement can be satisfied. M/G/R-PS model has been famous for estimating the application delay of the elastic data traffic. The M/G/R-PS model characterizes the TCP traffic at the flow level, where mobile users represent individual TCP flows generated by downloading Internet objects; the sojourn times represent the object transfer times. The transmission rate of UMTS user connections is limited by their assigned RAB type, e.g. 64 kbps.

The M/G/R-PS model has been studied in many researches to estimate the transfer delay of elastic data traffic on the flow level. The introduction of the basic M/G/R-PS model can be found in [14], and [15] proposes an extension of the basic M/G/R-PS model which considers the impact of TCP slow-start effect into the M/G/R-PS model, [16] applies the M/G/R-PS model for the UMTS network to estimate the application delay in case of single RAB service, either without or with call admission control. In this paper, for transmitting the elastic data traffic, we propose a new approach that is based on the M/G/R-PS model to dimension the IP-based Iub interface, where the above introduced DiffServ-based QoS structure is employed. In this QoS structured, the elastic traffic applications are accommodated by AF ore BE PHB services. Bandwidth allocation among different AF and BE PHB classes is implemented by the WFQ packet service discipline. If each AF/BE PHB queue is continuously sending data, then the kth queue receives the share of the available bandwidth BW_k as calculated in formula (1). That means, a fraction of the total capacity will be supposed to be allocated to each PHB. But if one PHB queue is inactive (i.e. not utilizing its allocated bandwidth), then its spare bandwidth shall be fairly shared among the other queues according to their weights.

Assuming each AF/BE PHB serves one NRT RAB service, the following parameters are defined:

- Let IP_BW be the total IP bandwidth;
- Let IP_BW_NRT be the total bandwidth allocated for NRT traffic, i.e. AF and BE PHBs;
- Let L_{EF} be the bandwidth occupied by the EF PHB;
- Let L_k be the mean offered load of the kth PHB of a AF or BE PHB;
- Let w_k be the weight of the kth PHB queue;
- Let r_k be the peak rate of the bearer service in the kth PHB queue;
- In case Call Admission Control (CAC) is in use, let s_k be the CAC guaranteed bit rate of the RAB service in the kth PHB queue;
- Let x_k be the average file size of the kth PHB queue.

The concept of the dimensioning for elastic traffic over AF or BE PHBs is illustrated in Fig. 4. The total IP bandwidth is divided into two parts: the bandwidth occupied by the RT traffic over the EF PHB, i.e. L_{EF} ; and the rest for IP_BW_NRT to transmit the NRT traffic over AF or BE PHBs. For each AF or BE PHB, the available bandwidth is the allocated BW for that PHB calculated with formula (1) adding the free bandwidth that is not utilized by other PHBs. The detailed steps of calculating the average application delay for the elastic traffic per AF or BE PHB are listed below.

1) The free capacity available for AF and BE PHBs is $IP\_BW\_NRT = (IP\_BW - L_{EF})$.
2) for the kth PHB, calculate the available capacity C_k with the following formula:

$$C_k = IP\_BW\_NRT \cdot \frac{w_k}{\sum_k w_k} + \left[IP\_BW\_NRT \cdot (1 - \frac{w_k}{\sum_k w_k}) - \sum_{j \neq k} L_j \right] \tag{2}$$

Here $IP\_BW\_NRT \cdot \frac{w_k}{\sum_k w_k} = (IP\_BW - L_{EF}) \cdot \frac{w_k}{\sum_k w_k}$ is the allocated bandwidth for own k^{th} PHB. It represents the minimum available IP bandwidth obtained by k^{th} PHB because all other PHBs are fully utilizing their bandwidth assigned by WFQ. $IP\_BW\_NRT \cdot (1 - \frac{w_k}{\sum_k w_k})$ gives the total allocated bandwidth for other AF and BE PHBs and $\sum_{j \neq k} L_j$ is the total offered traffic on other AF and BE PHBs, so $\left[IP\_BW\_NRT \cdot (1 - \frac{w_k}{\sum_k w_k}) - \sum_{j \neq k} L_j \right]$ equals to the spare bandwidth that is not utilized by other AF or BE PHBs. Therefore the available bandwidth C_k that can be used for the kth PHB is the sum of the theoretically allocated bandwidth for the kth PHB and the spare bandwidth that is not utilized by other PHBs.

3) for the kth PHB, calculate the offered load in percentage of the available bandwidth, $\rho_k = L_k / C_k$.
4) for the kth PHB with r_k (bearer rate), $R_k = C_k / r_k$.
5) If there is no limited connections for this PHB, i.e. when no admission control is applied, we calculate the average file transfer delay with the extended M/G/R-PS model in [16]. For the kth PHB, the calculation of the delay of transferring a file of length x_k with basic M/G/R-PS model is given below:

$$E_{M/G/R}\{T(x_k)\} = \frac{x_k}{r_k}\left(1 + \frac{E_2(R_k, R_k \rho_k)}{R_k(1-\rho_k)}\right) = \frac{x_k}{r_k} f_k \tag{3}$$

Where E_2 represents Erlang's second formula (Erlang C formula) and f_k is defined as the delay factor. The delay factor represents the increase of the average file transfer time (and decrease of the average throughput) due to the link congestion.

6) If applying CAC, there are limited N connections allowed
for this PHB, then in this case we should use the M/G/R/N-PS for calculating the average file transfer delay, where N is the maximum number of connections. For the RAB service in the kth PHB, the maximum num of connections of this RAB service is $N_k = (IP\_BW - L_{EF}) / s_k$. Probability of j customers in the system is given below [16]:

$$p(j)=\begin{cases}\dfrac{(1-\rho_k)\dfrac{R_k!}{j!}(R_k\rho_k)^{j-R_k}E_2(R_k,R_k\rho_k)}{1-E_2(R_k,R_k\rho_k)\rho_k^{\;N_k-R_k}\rho_k} & (j<R_k)\\[2ex] \dfrac{E_2(R_k,R_k\rho_k)\rho_k^{\;j-R_k}(1-\rho_k)}{1-E_2(R_k,R_k\rho_k)\rho_k^{\;N_k-R_k}\rho_k} & (R_k\le j\le N_k)\end{cases} \tag{4}$$

With the probability of each queue state, the average number of connections (or mean queue size) can be calculated. By applying Little's law, the average file transfer delay can be obtained by $E_{M/G/R/N-PS}\{T(x_k)\}=\dfrac{E\{W\}}{\lambda}$, where the average queue length is calculated with $E\{W\}=\sum_{j=0}^{N} j\cdot p(j)$.

It is noted that the above approach is for calculating the average application delay per AF or BE PHB. It is also applied to a single AF class with different drop levels. Because each drop level belongs to one specific PHB.

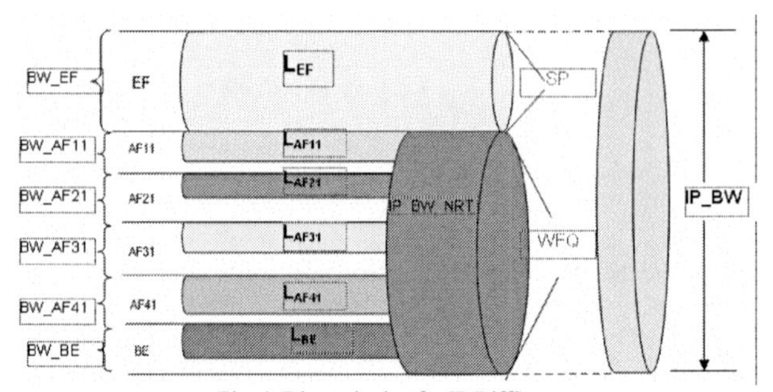

Fig. 4. Dimensioning for IP DiffServ

The above analytical approach is able to estimate the achieved average application performance in terms of mean transfer delay or application throughput for each AF or BE PHB. Given a particular application QoS requirement, from the proposed analytical model we can derive the total required IP bandwidth for the Iub link with the following steps:

(1) For each individual AF or BE PHB that serves a specific RAB service (e.g. RAB 384kbps), numerically calculate the required total IP bandwidth for carrying NRT traffic (i.e. *IP_BW_NRT*), which meets the maximum average transfer delay requirement for that RAB service, denoted as *IP_BW_NRT(service)*. The estimated average delay of the service on the kth PHB under certain *IP_BW_NRT* is computed with formula (2), (3) (without CAC) or (4) (with CAC). The required input is the mean offered traffic load on each PHB queue, then for a certain *IP_BW_NRT* value we can calculate the available bandwidth for service on the kth PHB C_k using formula (2). Then with C_k, we can derive $\rho_k = L_k / C_k$ and $R_k = C_k / r_k$ to calculate the average transfer delay in formula (3) or (4) depending whether there is admission control. We repeat the delay calculation numerically for a range of *IP_BW_NRT* values until the obtained delay value reaches the required delay boundary. Then we record the minimum *IP_BW_NRT* which meets the delay requirement as *IP_BW_NRT(service)*.

(2) Repeat the step (1) for all AF and BE PHBs. Then for every PHB each with a specific RAB service, there is one estimated *IP_BW_NRT(service)* that satisfies the delay boundary of that service.

(3) Take the maximum value of all *IP_BW_NRT(service)* to be the final *IP_BW_NRT*, because this is the worst case.

(4) Calculate the total required bandwidth on the Iub link (*IP_BW*). It is the sum of *IP_BW_NRT* and the bandwidth over EF PHB derived from the MD-Erlang model for RT streaming traffic. Additionally it may be added by the bandwidth reserved for DCCH, CCCH part.

6. VALIDATIONS

This section validates the proposed analytical dimensioning approach for the elastic traffic on the AF or BE PHB, to guarantee the application delay QoS of a service class by simulations.

6.1. SIMULATION SCENARIO

The validation is done by comparing with simulations. The simulation model is developed in OPNET, which include the IP DiffServ QoS structure. The simulation scenario consists of seven traffic types in UMTS. Table 1 gives the description of traffic types and their mapping to the DiffServ PHBs. In the investigated scenario, the CAC is applied for AF and BE PHBs allowing maximum 10 AF11, 10 AF21, 10 AF31, 6 AF41, 5 BE user connections, in addition there are 5 EF Video and 11 EF Voice UEs.

Traffic class	IP DiffServ PHB
Conversational (voice)	EF
Streaming (video)	EF
Interactive – R99 RAB 64kbps	AF11
Interactive – R99 RAB 128kbps	AF21
Interactive – R99 RAB 256kbps	AF31
Interactive – R99 RAB 384kbps	AF41
Background – HSPA (2Mbps peak rate)	BE

Table 1. Traffic Types and mapped PHBs

Their traffic models are described below. Table 2 gives the configuration of the WRED and WFQ parameters for AF and BE PHBs.

(1) Interactive Traffic Model – web application
- Inactive period (reading/thinking time period): geometric distribution (mean = 5s)
- File/Page size: constant distribution of 25kbyte

(2) Conversational Traffic Model – voice application
- AMR Codec
- Talking period: exponential distribution (mean = 3s)
- Silence period: exponential distribution (mean = 3s)
- Connection duration: exponential distribution (mean = 120s)
- Transport over UDP

(3) Streaming Traffic Model – video application (over UDP)
- 800bytes frame size
- 10 frames /second
- 64kbps coding rate

PHB	WRED Parameters				WFQ
	min_{th}	max_{th}	max_p	exp.w.	weight
AF11	4	10	10%	9	20
AF21	4	10	10%	9	30
AF31	5	10	10%	9	40
AF41	6	10	10%	9	50
BE	3	10	10%	9	10

Table 2. WRED and WFQ Parameters

The simulation scenario is composed of one NodeB and one RNC (in this work, we consider only single Iub scenario), and between them there are two IP DiffServ routers, which implements the DiffServ functions as explained in section 3.

6.2. VALIDATION RESULTS

As the CAC is used in this scenario for AF and BE PHBs, the M/G/R/N-PS model should be used to estimate the application delay of different RAB services over different PHBs. Fig. 5 compares the calculated application delay per PHB or RAB service using the M/G/R/N-PS model against the simulation results for the given scenario. It shows that for each PHB (AF or BE) and its serving NRT RAB service, given a certain Iub link utilization, the calculated application delay matches well with the delay values from the simulations. It proves that the proposed analytical model gives an accurate estimation of the application performance for various service classes each with different RAB service and priorities. And moreover, it is observed that when the Iub link is highly utilized, e.g. above 90% link utilization, the achieved delay performance is worse for the lower weight queue or service due to a less share of total bandwidth. For example, AF41 PHB has the highest weight so its delay performance is the best whereas BE PHB experiences the most delay degradation under the congestion.

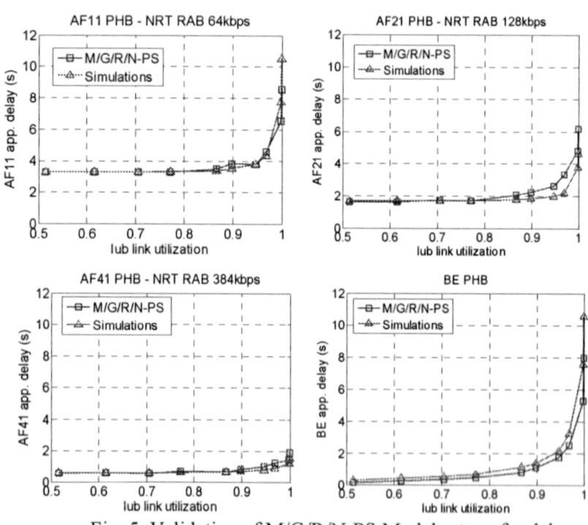

Fig. 5. Validation of M/G/R/N-PS Model – transfer delay

Fig. 6 presents for each PHB or RAB service the derived IP bandwidth allocated for all NRT traffic (i.e. *IP_BW_NRT*) using the M/G/R/N-PS model under different application delay requirements of that RAB service, and compares them against the simulation results. It shows that for all RAB services, when the delay QoS requirement is higher, i.e. the desired application delay need to be smaller, then a higher IP bandwidth is required. As explained in the above dimensioning steps in section V, if each RAB service has a specific delay requirement, then for each service we can calculate the required total IP bandwidth for all NRT traffic (i.e. *IP_BW_NRT*) under which the delay boundary of that service is satisfied, and we should take the maximum of the calculated bandwidth as the total IP bandwidth for carrying the NRT traffic (i.e. *IP_BW_NRT*). By comparing the simulation results and the derived IP bandwidth from the proposed model in Fig. 6, it shows that in general the proposed analytical approach based on the M/G/R/N-PS model can give a relatively good approximation of the required IP bandwidth for the dimensioning results, so it is proven to be a proper dimensioning approach.

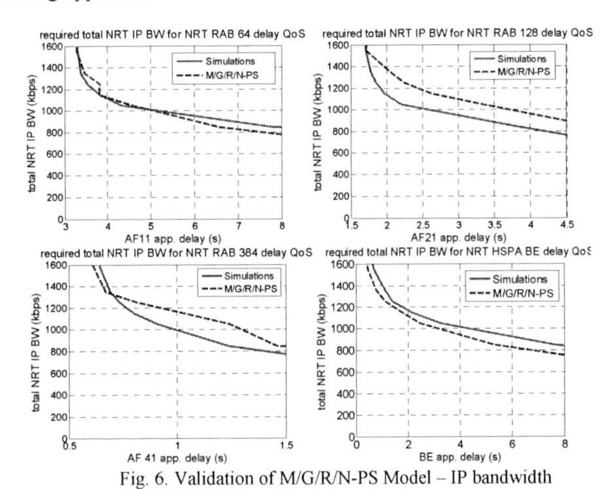

Fig. 6. Validation of M/G/R/N-PS Model – IP bandwidth

7. IMPACT OF DIFFSERV ON IUB DIMENSIONING

This section analyzes the impact of using DiffServ QoS scheme on the Iub dimensioning. Here we compare the performance with two different schedulers in DiffServ Router. (I) only Strict Priority; (II) combined SP & WFQ as explained in section 4. The structure of (I) is shown in Fig. 7.

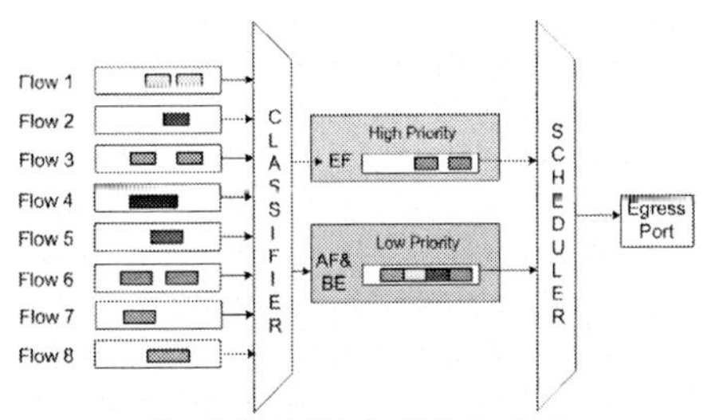

Fig. 7. Strict Priority QoS structure

In structure (I) both AF and BE PHB traffic are going to one shared queue, whereas in (II) there is one dedicated queue for each AF or BE PHB and among them a WFQ scheduler is used. Moreover, we try two different WFQ weight settings for the structure (II). The configurations of the WFQ weight for different PHBs are given in Table 3, named as DiffServ 1 and DiffServ 2 respectively.

PHB	DiffServ 1 WFQ Weight	DiffServ 2 WFQ Weight
AF11	20	50
AF21	30	40
AF31	40	20
AF41	50	20
BE	10	10

Table 3. Two WFQ settings for the combined SP & WFQ

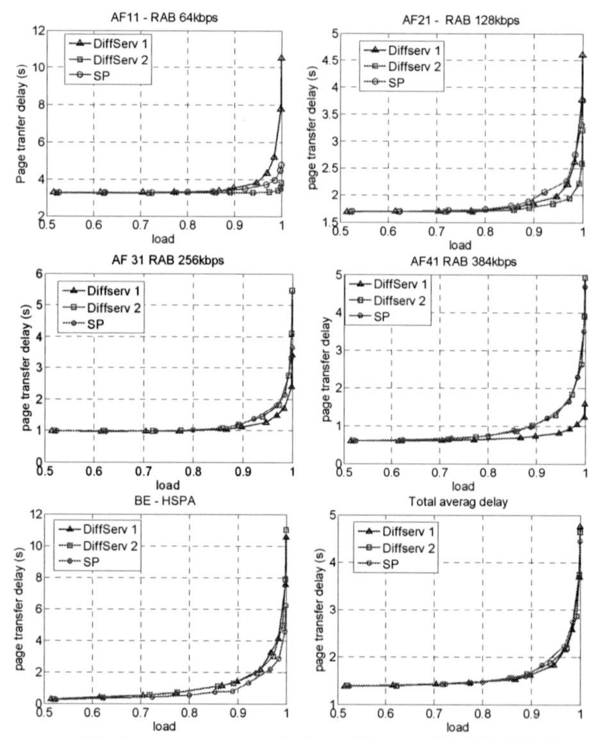

Fig. 8. average page transfer delay (SP vs. combined SP & WFQ

Fig. 8 gives the comparison of the achieved average page transfer delay at the user application layer per AF/BE PHB, using (I) SP, and (II) with DiffServ 1 and DiffServ 2 settings respectively. From this group of results, we can conclude that using the queue structure (II), i.e. combined SP and WFQ, the delay performance for lower weight PHBs is worse than SP; the delay performance for higher weight PHBs is improved. But for the total average delay of all PHBs, IP DiffServ and SP achieve similar results. That means, the scheduler does not have significant impact on the total average delay. But the use of WFQ scheduler allows setting different priorities for different traffic classes, and then the traffic can be treated differently by the network: the higher priority traffic class is given a higher weight in the WFQ scheduler and thus its delay performance is better guaranteed.

8. CONCLUSION

In this paper, we proposed a sophisticated QoS structure in the IP-based UTRAN, which is based on DiffServ QoS scheme with an integrated Weighted Fair Queue (WFQ) and Strict Priority (SP) scheduling. According to this QoS structure, we proposed analytical dimensioning approaches for deriving the required IP Bandwidth on the Iub interface, for elastic NRT traffic and RT streaming individually. The proposed dimensioning approach for elastic traffic is based on the extension work on the M/G/R-PS model; the analytical approach was validated with simulations. Moreover, we analyzed the impact of IP DiffServ on the dimensioning by comparing performance against Strict Priority Scheduler.

REFERENCES

[1] 3GPP TS 25.855, 3rd Generation Partnership Project; Technical Specification Group Radio Access Network, High Speed Downlink Packet Access (HSDPA): Overall UTRAN description (Release 5).

[2] 3GPP 3GPP TS 25.309 V6.3.0, 3rd Generation Partnership Project; Technical Specification Group Radio Access Network, FDD Enhanced Uplink; Overall description (Release 6).

[3] 3GPP, *IP Transport in UTRAN Work Task Technical Report*, TR 25.933, TSG RAN WG3.

[4] MOBILE WIRELESS INTERNET FORUM (MWIF), *IP in the RAN as a Transport Option in 3rd Generation Mobile Systems*, Technical Report MTR-006, Rel. V2.0.0, June 18, 2001

[5] S. BLAKE, D. BLACK, M. CARLSON, E. DAVIES, Z. WANG, AND W. WEISS, *An architecture for differentiated services. Request for Comments (Informational) 2475*, IETF, December 1998.

[6] R. BRADEN, D. CLARK, AND S. SHENKER, *Integrated services in the internet architecture: an overview. Request for Comments (Informational) 1633*, Internet Engineering Task Force, June 1994

[7] R. BRADEN, L. ZHANG, S. BERSON, S. HERZOG, AND S. JAMIN, *Resource reservation protocol (RSVP), version 1 functional specification, Request for Comments (Standards Track) 2205*, IETF, September 1997

[8] E. ROSEN, A. VISWANATHAN, AND R. CALLON, *Multiprotocol label switching architecture. Request for Comments (Proposed Standard) 3031*, Internet Engineering Task Force, January 2001.

[9] JUKKA MANNER, *Provision of Quality of Service in IP-based Mobile Access Networks*, Ph.D. Thesis, University of Helsinki, Department of Computer Science, November 2003, Report number A-2003-8.

[10] F. AGHAREBPARAST AND V.C.M. LEUNG, *QoS support in the UMTS/GPRS backbone network using DiffServ*, IEEE Global Telecommunications Conference, vol.2, pp.1440–1444, Nov. 2002.

[11] NICHOLS, K., BLAKE, S., BAKER, F. AND D. BLACK, *Definition of the Differentiated Services Field (DS Field) in the IPv4 and IPv6 Headers*, RFC 2474, December 1998.

[12] S. FLOYD AND V. JACOBSON, "*Random early detection gateway for congestion avoidance*," IEEE/ACM Trans. Networking, vol. 1, pp. 397–413, August 1993.

[13] A. DEMERS, S. KESHAV, AND S. SHENKER, *Analysis and simulation of a fair Queueing algorithm*, J. Internetworking Res. Experience, pp. 3–26, Oct. 1990; also in Proc. ACM SIGCOMM'89, pp. 3–12.

[14] ZHONG FAN, *Dimensioning Bandwidth for Elastic Traffic*, Networking 2002, pp 826-837.

[15] ANTON RIEDL; THOMAS BAUSCHERT; MAREN PERSKE; ANDREAS PROBST; *Investigation of the M/G/R Processor Sharing Model for Dimensioning of IP Access Networks with Elastic Traffic*, First Polish-German Teletraffic Symposium PGTS 2000,Dresden, September 2000.

[16] X. LI, R. SCHELB, C. GÖRG AND A. TIMM-GIEL, *Dimensioning of UTRAN Iub Links for Elastic Internet Traffic*, in Proc. 19th International Teletraffic Congress, Beijing, Aug/Sept. 2005, 2005.

Web sites traffic analysis, measurement, locality, AS, BGP prefixes

Joachim CHARZINSKI*

LOCALITY ANALYSIS OF TODAY'S INTERNET WEB SERVICES

In the beginning of the Web, one Web page was equal to one HTML file that was retrieved from one server and contained links to other pages which might reside on other servers. Today, Web pages contain many elements that are collected from different servers. With the advent of Web APIs, even mashups are designed that retrieve and combine data from a number of remote servers on the Internet. This paper investigates the locality structure of modern Internet Web services by analyzing the number of hosts, routing domains and external services contacted by a Web client that uses some of today's most popular Web sites. Those results are compared to the characteristics shown by selected mashup sites.

1. INTRODUCTION

During the first years of the Web, web sites had a high degree of locality. One Web page was equal to one HTML (Hypertext markup language) file that could be retrieved from one server. Those pages contained textual information and links to other pages that might reside on a different server and could be retrieved next. Soon after, web pages and web sites started getting more distributed and more dynamic. Images were embedded on Web pages, pages were constructed dynamically and could contain multiple sub-pages (frames) from different sources. Nowadays, a large number of Web sites contact a high number of different servers to retrieve images, page elements or even raw data, as in the case of *Web 2.0 mashups* that utilize APIs (application program interfaces) provided by other services to construct new value-added services. In addition to plain HTML pages, developers now extensively include JavaScript and Flash code on the pages that retrieve additional elements from Web servers either autonomously or triggered by mouse events or timers. Correspondingly, not all application behavior is related to human activity any more. Also, especially the JavaScript and Flash contents makes it impossible to predict Web pages' traffic behavior from a static code analysis. Instead, applications can only be analyzed decently by tracing actual Web usage.

This paper investigates the location properties of some of the most popular web sites in terms of the number of different hosts, the number of different routing domains and the number of different organizations contacted. By comparing the world's most popular Web sites with some selected mashups, it can be seen that a high degree of distribution has already become "normal" on the Web.

*Nokia Siemens Networks, Munich, Germany, j.charzinski@ieee.org

1.1. RELATED WORK

Crovella and Krishnamurthy [5] have written a comprehensive book on Internet Measurement, including passive and active measurement approaches and the most important findings in traffic characterization. Traffic characteristics and topology properties of new applications like peer-to-peer based services have been analyzed by numerous authors, e.g. in [11], but to the author's knowledge, there is no measurement study on the locality of Internet Web services and mash-up traffic yet.

Geographic locality has been studied recently in some publications. Fonseca et al [6] focused on caching and location of proxies. Lakhina et al [7] study the geographic locations of Internet nodes, Autonomous Systems and links. Spring et al [10] give a good overview of methods to analyze the Internet's network structure and topology.

This paper in contrast focuses on application properties and logical locations such as domain ownership and routing towards the different servers that make up a single Internet service such as *google maps, weather.com* or *myspace.com*.

1.2. OVERVIEW OF THE PAPER

Sec. 2 summarizes some important aspects in Web evolution. The methodology used to obtain the results presented in this paper is introduced in Sec. 3 and results are given in Sec. 4 both for the number of external services that a Web site *uses* and for the number of external Web sites that a service is *used by*.

2. EVOLUTION OF WEB APPLICATIONS

The phenomenon of *mashups* [12], i.e. the re-combination of data from various Web sites into a new service, is the culmination of a trend that has been going on for several years. Internet Web sites have long evolved to include elements such as images or frames from other Web sites. The new development of mashups is that not only elements are retrieved from other sites that can be directly included in a Web page and displayed by a browser, but also the direct access to raw data is offered and used. Early – and mostly unofficial – mashup sites used screen scraping [9] to extract data from other Web sites whereas nowadays those Web sites offer APIs (application program interfaces) which allow direct access to the raw data and even provide advanced data access methods such as complex queries on the data or authentication to the service.

This approach seen on the Internet has some commonality with but is still distinctly different from the Enterprise software architectures described as Web Services or SOA (Service Oriented Architecture). Both share the idea of using one service's data in another service, but whereas in the Enterprise world the focus is on structuring and separating applications, the focus in the Internet world is on re-using data that have a value of their own in the original services for building new services. Also, the Internet world sees a lot more freedom of standards, e.g., a number of different methods for addressing APIs such as REST (Representational State Transfer) versus SOAP (Simple Object Access Protocol).

The input data obtained via an API can be combined by applications running on the Web client, such as Flash, JavaScript or Java applications, (see Fig. 1 left), or by applications running on the Web server(s) (see Fig. 1 right). Depending on the architecture chosen for a mashup, the access to the original source of data can or cannot be seen at the client side when using the mashup service.

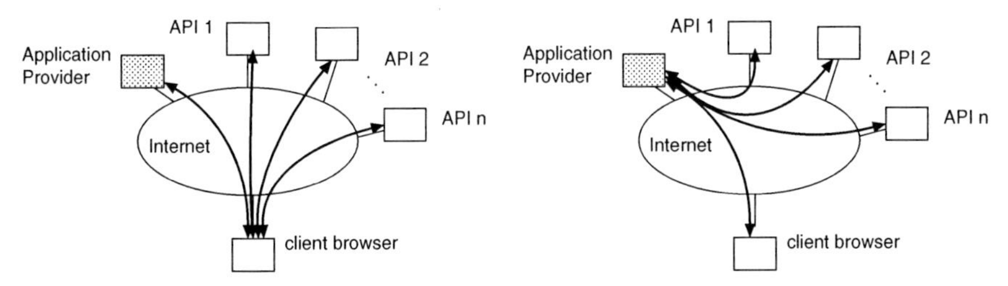

Fig. 1: Principal connectivity architecture of client (left) and server side mashups (right).

A large number of Web sites is already using and/or offering mashups APIs. Fig. 2 shows an example of a rank statistic for Web APIs used by mashups as observed by the programmableweb site [3].

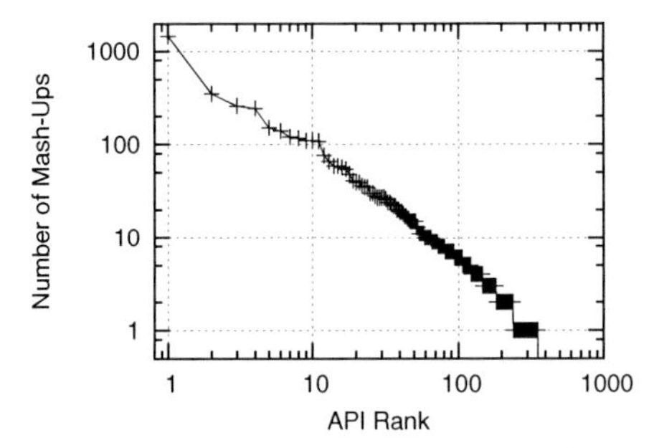

Fig. 2: Ranking of APIs by number of mashups using the respective API. Data from [3]

It can be seen that out of the more than 300 APIs counted by [3], only a few (roughly 10) are used by more than 100 services while most of them are used by less than ten sites.

3. METHODOLOGY

The evaluations in Sec. 4 are performed on client-side measurements of Internet traffic, i.e. all the traffic from and to a client machine is observed, but the traffic between different Internet servers cannot be seen. Correspondingly, the servers involved in client-side mashups can be investigated, but it cannot be determined what traffic is generated on the Internet due to server-side mashups (cf. Fig. 1).

3.1. LOCALITY

There are different logical views on locality to explore, depending on what aspect of networking is in focus. As sketched in Fig. 3, evaluations can be performed on

a. **Hosts**, determining how many different hosts are contacted to retrieve all the data for a Web page. This is relevant for a (TCP) connection level view on traffic, as required e.g. for per-connection QoS reservations or to determine firewall rules.

b. **Routing Domains**, looking at which target networks the contacted hosts are residing in. This is relevant e.g. for a network peering view or to determine firewall rules.

c. **Organizations**, focusing on the logical mapping to organizations providing a service. This is relevant e.g. when judging trust or legal relations.

d. **Geography**, looking at the geographical locations of the servers contacted. As with c., this can be relevant when judging legal relations, e.g., the trustworthiness of jurisdictions the servers are operated in, but also to determine the minimum possible packet latencies.

This paper focuses on the first three aspects. Routing domains are assessed both in terms of BGP core routing table prefixes and Autonomous System Nubers (ASNs) for the observed hosts. DNS second level domains (SLDs) are used as an approximation to analyzing organizations.

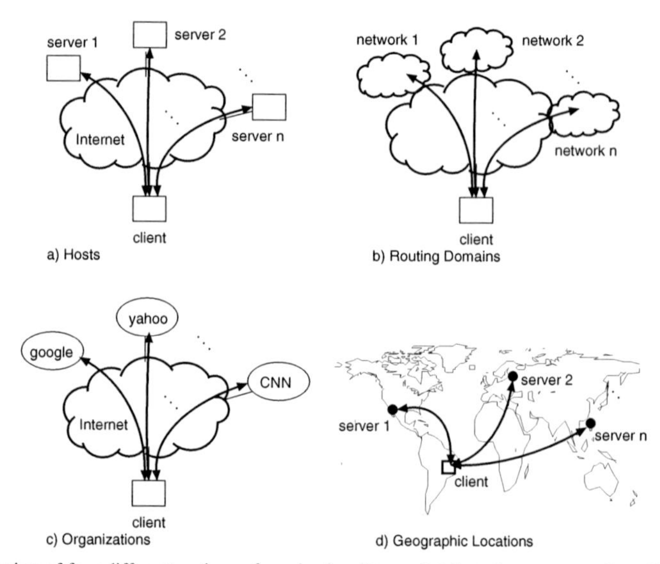

Fig. 3: Illustration of four different notions of service locality or distributedness, as seen by a Web client, e.g., in a client-side mashup.

3.2. DIRECTED EXPERIMENTS

There are two distinctly different ways of determining traffic characteristics: **passive observations** evaluate the traffic caused by a large number of Internet users at an Internet access network has the advantage of representing users' browsing habits and preferences as well as yielding good statistics due to a high number of samples. On the other hand, it is tricky to extract per-site or per-click data from passively recorded packet traces and the heuristics that have to be used (cf. e.g. the one-click heuristic in [4]) are always disputable. Privacy issues also often prevent researchers from analyzing the services actually used by real users. Similarly, evaluating traffic at Web servers gives good insight into the incoming and outgoing traffic for the respective server, but neglects client browsing preferences. In addition, the access to server based information is hardly possible for the most relevant Internet sites.

Active observations initiate and control the observed Web session as part of the experiment. In this way, after selecting the sites to be visited, the observed traffic can be fully attributed to the corresponding site. However, care must be taken to ensure sensible browsing behavior while performing the measurements. Also, automation is restricted to basic site selection, but when representative usage of a Web site is required, an automated crawler cannot distinguish between a popular and an unpopular link on a site. Additionally, it is hard to implement crawlers that use the Flash applications found on many of today's Web sites and to emulate the actions of JavaScript elements such as downloads triggered by "mouseover" events in automated systems.

The measurements evaluated in this paper fall into the group of active observations. Two groups of data sets were generated in May and June 2008 by first selecting sites to be visited and then browsing each of the sites for about 2–5 minutes each, using most of the site's features but only selecting links that stay on the site, i.e., avoiding links that would lead to leaving the site and continuing on another Web site.

The sites to be visited have been selected from two lists: The first data set, "**Web top 25**" are the top 25 (actually, skipping one adult service and including number 26 instead) US sites most popular among users as listed by the Alexa [1] Web statistics service. These include a mixture of search, news, trading, contents sharing and social networking sites. Additionally, there were two sessions recorded for the most popular site, google.com. In a first session ("search only"), only the main (text) search page was utilized, without following external links. In a second session ("full"), all the different search options offered by google were utilized, i.e., the search for text, images, maps, photos, ... The second data set, "**mashups**" are the top 18 mashup sites as listed by programmableweb [3]. This list has been chosen mainly for the lack of more reliable classification of Web sites.

A `tcpdump` [8] process on the Web client machine recorded all packet headers and after the sessions the following parameters were evaluated:

- number of HTTP/TCP connections established, evaluated through TCP SYN packets seen in the traces
- number of different servers contacted, evaluated from the destination IP addresses of TCP SYN packets
- number of different routing domains contacted, evaluated by performing a longest prefix match on the TCP SYN packet destination IP addresses in the route-views BGP table from [2]
- number of different DNS domains contacted, evaluated by analyzing the DNS lookups recorded in the trace as well as by determining the responsible DNS server (DNS NS record) for the routing domains seen above. Only the last two top-level and second level domain names have been taken into consideration. E.g., all of "photobucket.com", "i90.photobucket.com" and "img.photobucket.com" are considered the same DNS domain.

4. EVALUATIONS

Sec. 4.1 evaluates *active* usage, i.e., what external services a specific site uses. *Passive* usage covers the sites a specific service is used by, see Sec. 4.2. Note that the exact numbers depend on the usage within the sessions. They should be seen as indicating tendencies but not as generally correct statements about the services.

4.1. ACTIVE USAGE

Hosts

Fig. 4 shows the rank distribution for the number of hosts contacted per site visit. Most Web sites are distributed applications that require the browser to contact many (up to 100) different servers.

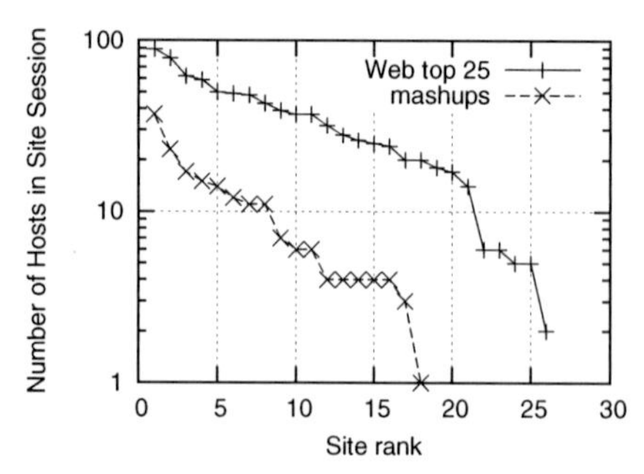

Fig. 4: Rank distribution of number of hosts contacted per site visit.

A compensation effect between the number of hosts contacted during a visit to a Web site and the number of connections established to each of those hosts could be expected. This is explored in the scatter plots in Fig. 5 which plot the average number of connections to each host versus the number of hosts contacted per site session. Each point in the plot corresponds to one trace file, i.e., one site session. It can be seen that the number of connections to each server neither increases nor decreases with the number of servers contacted.

Routing Domains

Two notions of routing domains are used in this paper:

1. BGP core routing table prefixes signify different entries in the core BGP routing table seen in [2].

2. Autonomous system numbers (**ASN**s) signify the autonomous systems (the network) which the prefixes point to. Due to the fragmentation of BGP tables, there can be multiple routing table entries with different prefixes for the same ASN.

Fig. 6 gives the rank distribution of the number of different prefixes (left) and ASNs (right) that the hosts contacted during a site visit belong to. If a network provider wanted to enhance users' experience with the top 25 Web sites and ensure sufficient capacity for those services, reservations

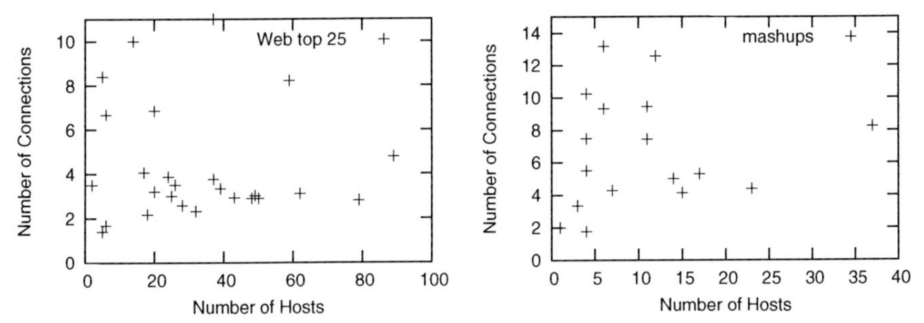

Fig. 5: Scatter plot of number of connections per host versus number of hosts contacted when visiting the Web top 25 (left) or mashup sites (right).

would have to be made for each of the routes to the networks seen here. Note that on average there are 1.5 different prefixes contacted per AS network.

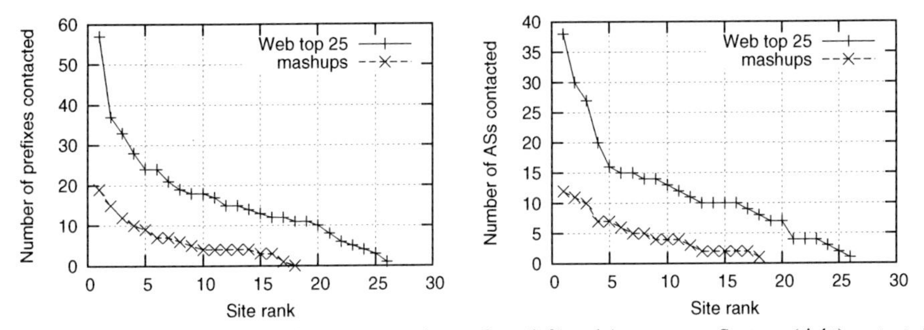

Fig. 6: Rank statistics of number of BGP core routing prefixes (left) and Autonomous Systems (right) contacted per site.

The plots in Fig. 7 show that on average 1–3 hosts are contacted in each network prefix. Obviously, the number of hosts contacted is a lower limit to the number of routing domains.

Comparing the plots in Fig. 7, we see that there is no significant difference between the locality of the top 25 Web sites and specially selected mashup sites. Both show two groups of sites, one contacting only a few (one to ten) routing domains and hosts and one contacting between ten and 100 hosts and routing domains. Fig. 8 shows again that there are mostly between one and two network prefixes addressed per AS. This figure is only given for the "Web top 25" services, but results are similar for the top mashup sites.

DNS Domains

While the above routing domain analysis shows how complex the configurations would be for reserving capacity or setting up firewalls that support access to the top Web sites, it over-estimates the structural complexity of the services. When a site uses an API, there is no big difference if the API

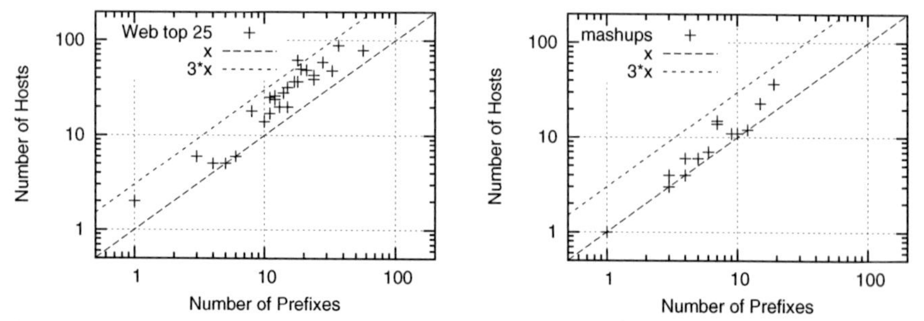

Fig. 7: Scatter plot of number of hosts versus number of network prefixes contacted in Web top 25 (left) or top mashup sites (right).

provider performs load balancing or geography matching by directing traffic to many routing domains in the world or keeping all connections to a single routing domain.

Therefore this section investigates the number of different Web service providers contacted during a client's visit to a Web site. As a heuristic, we consider DNS domains (analyzed as described in Sec. 3.2) equivalent to Web service providers, accepting the limitations that this approach neglects that (1) some hosting providers serve a number of Web service providers and (2) some Web service providers offer a number of fundamentally different services from the same DNS second-level domain (SLD).

The rank distribution of the number of DNS SLDs contacted per site is depicted in Fig. 9. Similar to the number of hosts, network prefixes or ASs contacted (see Figs. 4 and 6, there are some sites that make browsers contact a large number (more than 30) of services and some that remain local (only one DNS SLD) or contact only few external services.

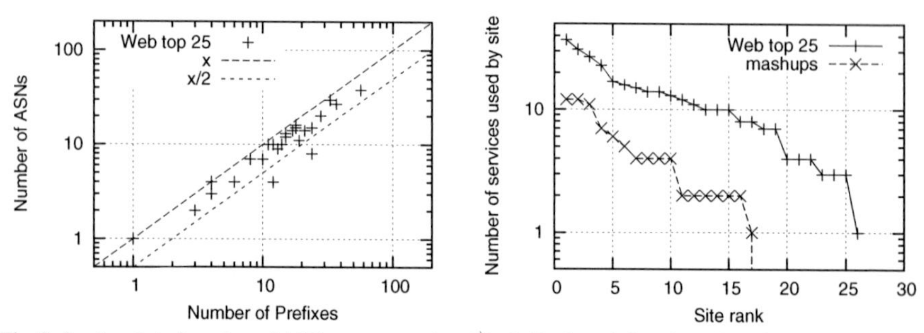

Fig. 8: Scatter plot of number of ASNs versus number of routing prefixes contacted by Web top 25 sites.

Fig. 9: Rank statistics of number of DNS domains contacted per site session.

The difference between the "Web top 25" and "mashups" distributions stems from the fact that the mashups sites featured on [3] are not as sophisticated as the top 25 Web sites on the Internet and therefore e.g. do not contact as many advertisement banner sites.

Tables 1 and 2 summarize the observed characteristics for the most heavily distributed (Tab. 1) and the most local of the top 25 Web sites (Tab. 2). Note that the number of connections has been indicated here for reference only. Much more than the number of hosts, DNS SLDs, network prefixes or ASs contacted, the number of connections seen for some sites depends heavily on the duration of a client session and on the client activity during the session. Both tables have been sorted according to the hosts column.

Table 1: Sites using the highest number of external services. (NP = network prefix, ASN = Autonomous System Number, DD = DNS domain, mHpASN = maximum number of hosts per ASN)

Site	Hosts	NPs	ASNs	DDs	mHpASN	Conn.
http://photobucket.com	89	37	27	35	28	425
http://blogger.com	79	57	38	44	14	222
http://www.myspace.com	62	18	16	17	20	192
http://weather.com	59	28	20	24	9	485
http://aol.com	50	19	11	15	19	145
http://aim.com	49	21	14	16	15	148
http://wordpress.com	48	33	30	29	4	137
http://www.google.com (full)	43	24	8	8	36	120
http://msn.com	39	24	15	19	14	130
http://www.yahoo.com	37	17	14	9	9	139

Table 2: Sites using fewest external services. (NP = network prefix, ASN = Autonomous System Number, DD = DNS domain, mHpASN = maximum number of hosts per ASN)

Site	Hosts	NPs	ASNs	DDs	mHpASN	Conn.
http://wikipedia.org	2	1	1	1	2	7
http://craigslist.org	4	4	4	3	1	36
http://facebook.com	5	4	3	3	2	7
http://www.google.com (search only)	6	6	4	4	3	10
http://rapidshare.com	6	3	2	3	5	40

Except for wikipedia.org, all observed sites make the browser include elements or data from external services.

The plots in Fig. 10 give scatter plots visualizing the number of network prefixes contacted versus the number of DNS SLDs contacted when visiting a site. The average number of network prefixes contacted per service used (as indicated by DNS SLD) is again between one and three. Except for the (1,1) sample, all sites make the browser include data from at least three different services. The more services are included by a site, the more single highly distributed services are averaged out in the overall count, so that the largest average number of network prefixes per DNS SLD is observed with sites including three to eight different services.

In line with the previous discussion, the results for selected mashups are similar to those for the Web top 25 sites, but there are fewer highly distributed sites among the selected mashups than among the top Web sites.

4.2. PASSIVE USAGE

In contrast to the previous section, now the other side of the data delivery chain is investigated using the same data sets as before but evaluating how many sites make the browser contact specific services. Fig. 12 depicts the rank distribution of the most contacted network prefixes, which is taken as an approximation for the most popular services to be included in other sites.

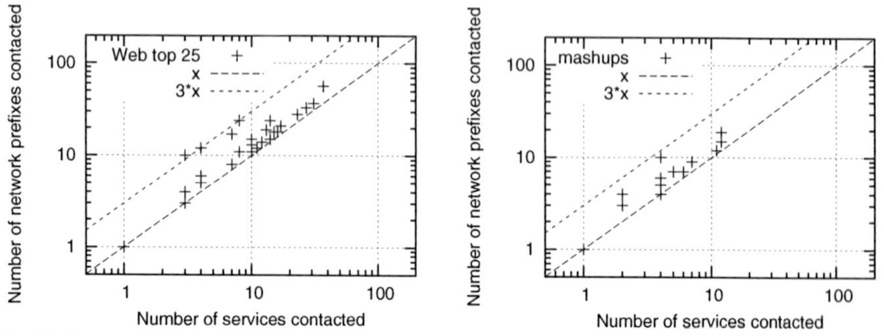

Fig. 10: Scatter plot of number of network prefixes vs. number of DNS domains contacted by top 25 Web (left) or top mashup services (right).

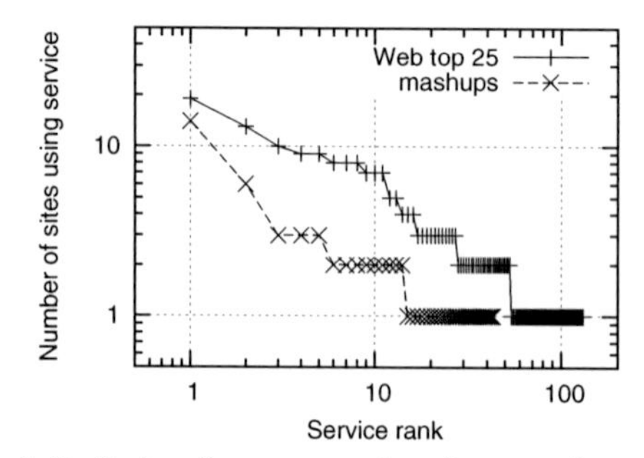

Fig. 11: Rank distribution of most contacted services according to DNS SLDs.

Tables 3 and 4 summarize the characteristic data for the services most included by the top 25 Web sites (Tab. 3) and the selected mashups (Tab. 4). In both tables, the search and service providers Google and Yahoo are important players. In addition, the – mostly commercial – Web top 25 sites offload most of their more static contents to content distribution networks such as Akamai in order to bring the contents close to end users. The measurements in this paper were done from the Arcor DSL network in Germany, and the Akamai hosts were seen as being part of the Arcor network. Server hosting (Rackspace, Theplanet) is utilized by both classes of Web sites. The main difference between the Web top 25 and more experimental mashup services is in the additional external services they use beyond Google and Yahoo. Many of the top Web sites use Doubleclick's advertisement referal services as a source of additional revenues. The explicit mashup sites employ a more diverse bouquet of external services, and Amazon is one example that was used by three of the investigated sites. Although many of the top sites are using akamai.net and doubleclick.net, they are not making clients use an API but include Web elements (images or frames) that are retrieved from those services, i.e., not mashups in the stronger sense.

When interpreting the number of prefixes and the number of ASNs, it is important to note that an Internet router can have different routes to different prefixes of the same AS. For instance, the Google network (AS 15169) is listed with 131 different network prefixes in the OIX BGP table [2] and evaluating the best AS paths yields three groups of prefixes with three different next hops for the

Google AS. Similarly, the Doubleclick network (AS 6432) consists of 44 network prefixes with 13 different best AS paths and 5 different next hops seen from OIX. Note that this will be different in different places of the Internet core.

Table 3: Services most used by other sites out of the Web top 25 list and the number of network prefixes (NPs) and Autonomous System Numbers (ASNs) seen for those services. (Note that one of the top 25 sites is google, which was visited in two different sessions, a search-only and a full service session, leading to the total number of session samples being 26 for the top 25 Web sites)

Service	sites using this service	NPs seen for service	ASNs seen for service
akamai.net	21/26	4	4
google.com	19/26	22	1
doubleclick.net	13/26	7	1
rackspace.com	9/26	4	2
yahoo.com	9/26	18	9

Table 4: Services most used by other sites out of the Web top 25 list and the number of network prefixes (NPs) and Autonomous System Numbers (ASNs) seen for those services.

Service	sites using this service	NPs seen for service	ASNs seen for service
google.com	15/18	13	1
yahoo.com	6/18	8	3
akamai.net	6/18	1	1
amazon.com	3/18	2	1
theplanet.com	3/18	3	1

Looking again at network prefixes in addition to DNS SLDs, Fig. 12 shows the rank distribution of the number of sites contacting a network prefix. This distribution is highly similar to the distribution of the number of sites contacting a service, but not identical to it.

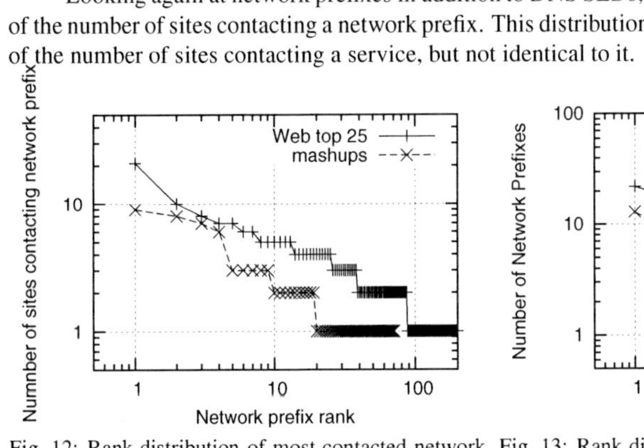

Fig. 12: Rank distribution of most contacted network prefixes.

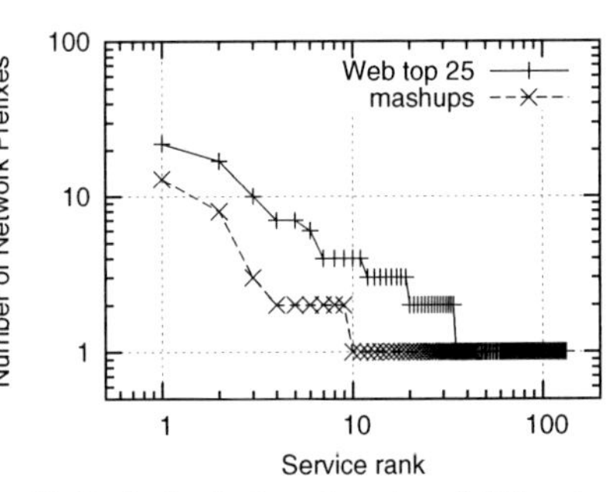

Fig. 13: Rank distribution of number of different network prefixes contacted per service used by other sites.

In Fig. 13, the number of network prefixes seen per service (classified by DNS SLDs) is plotted. Although the figure seems very similar to the previous ones, it shows something completely different, namely the second column of Tables 3 and 4. A *single* service (google.com) that is used by other sites was contacted at more than 20 different network prefixes, showing again the high degree of distribution google has implemented with their servers.

At the other end of the extreme, there are many services which are only contacted through one network prefix. Investigating this relationship further, Fig. 14 shows a scatter plot of the number of network prefixes seen for a service versus the number of sites using this service. There is a clear correlation showing that it is the most used services that also have the highest degree of distribution with respect to network prefixes. On the other hand, some of the less used services still are contacted in multiple network prefixes.

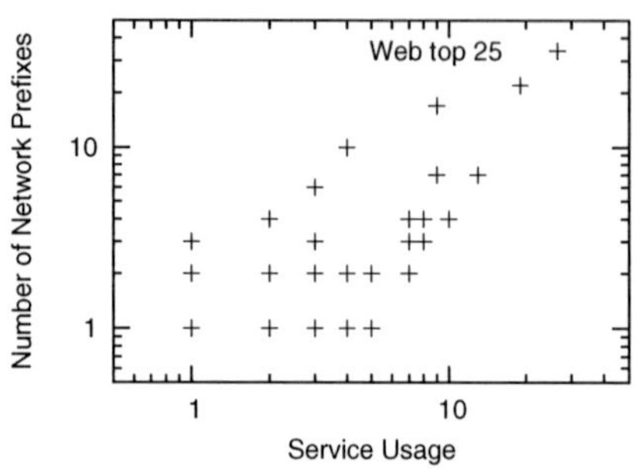

Fig. 14: Scatter plot of number of network prefixes contacted per service versus number of sites using a service.

5. CONCLUSIONS

In this paper, a number of top Web sites have been investigated with respect to the number of additional external services that are contacted by clients to complete web pages when visiting the sites. A directed experiment approach was chosen to allow per-site, per-DNS-domain and per-routing-domain analyses that would have been impossible with passive traces anonymized to comply with privacy requirements.

Not only the Internet as a transport infrastructure but also the Internet as a service structure has become a highly distributed construct where Web sites heavily employ external services, be it in the form of embedded images, HTML frames or raw data accessed through APIs.

The study found that there is no significant structural difference in the locality of the client traffic caused by explicit mashup sites (sites that use raw data from other sites via APIs) and the traffic observed when visiting some of the most popular Web sites. Most of the popular sites extensively employ data or elements from external services. Some of these services, especially the more popular ones, are again widely distributed, as can be seen in the analysis of the number of different hosts, network prefixes and autonomous systems that these services were contacted through.

The high degree of distribution of today's Internet Web services has a number of consequences for Internet transport, service, and security design:

The **availability** ratio of a compound service is only the product of the availability ratios of its independent components. This makes compound services much more vulnerable to server outages as well as routing problems to some parts of the Internet than would be the case for services with stronger locality.

The high degree of distribution makes it hard to include today's top Internet Web services in **security** designs that filter traffic on the basis of host-to-host relations. Whereas this might not be an issue for the open Internet, the emerging trend to use the mashup paradigm also for Enterprise Web services (including services from the open Internet) might turn into a problem for IT (Information Technology) departments.

Providing QoS (Quality of Service) support for the most relevant Internet Web services is getting more and more difficult as the degree of distribution increases. In order to provide high capacity connectivity for e.g. the full set of services from Google alone, an Internet access provider would have to provide dedicated capacity to more than 20 different network prefixes. Supporting Yahoo services in addition would require some further 20 network prefixes to support capacity reservations on. Reserving capacity for a single Web session becomes completely infeasible in this way, but even static reservations between access networks and Internet services would require an unrealistically high management effort. Consequently, the decreasing locality of Internet Web services further supports the current Internet capacity design paradigm where transport providers increase their capacity to support unspecified services and service providers optimize their services to be close enough to the most relevant transport providers to support a satisfying user experience.

In order to get a more representative view of Internet services, a number of extensions of the study will be required. More relevant sites have to be sampled to get a better statistical basis of the data. The study would have to be repeated with clients in different access networks to allow comparing results to better assess the effect of content distribution networks. Services with multiple second level domain names have to be consolidated, e.g. ebay with ebaydns, ebayimg and ebayrtm, in order to get a correct view on the organizational structure behind the services contacted when visiting a site.

[1] Alexa: most popular US web sites. available at http://www.alexa.com/site/ds/top_sites?cc=US&ts_mode=country&lang=none, visited June 16, 2008.

[2] OIX Route Views. http://archive.routeviews.org/oix-route-views/2008.05/, visited 28. May 2008.

[3] Programmable Web. http://www.programmableweb.com/, visited June 11, 2008.

[4] J. Charzinski. Observed Performance of Elastic Internet Applications. *Computer Communication*, 26(8):914–925, 2003.

[5] Mark Crovella and Balachander Krishnamurthy. *Internet Measurement: Infrastructure, Traffic and Applications*. John Wiley and Sons, Inc, 2006.

[6] Rodrigo Fonseca, Virgílio Almeida, and Mark Crovella. Locality in a web of streams. *Communications of the ACM*, 48(1):82–88, January 2005.

[7] Anukool Lakhina, John W. Byers, Mark Crovella, and Ibrahim Matta. On the geographic location of internet resources. *IEEE Journal on Selected Areas in Communications, Special Issue on Internet and WWW Measurement, Mapping, and Modeling*, 2003.

[8] S. McCanne, C. Leres, and V. Jacobson. tcpdump. LBNL Network Research Group, available at ftp://ftp.ee.lbl.gov/tcpdump.tar.Z.

[9] Michael Schrenk. *Webbots, Spiders, and Screen Scrapers*. No Starch Press, 2007.

[10] Neil Spring, David Wetherall, and Thomas Anderson. Reverse engineering the internet. In *2nd workshop on hot topics in networks*, Nov. 2003.

[11] Kurt Tutschku and Phuoc Tran-Gia. Traffic characteristics and performance evaluation of peer-to-peer systems. In Klaus Wehrle Ralf Steinmetz, editor, *Peer-to-Peer-Systems and Applications*. Springer, 2005.

[12] Raymond Yee. *Pro Web 2.0 Mashups: Remixing Data and Web Services*. Apress, 2008.

III. INVITED TALKS

Petter HOLME*

MODELING THE EVOLUTION OF THE AS-LEVEL INTERNET: INTEGRATING ASPECTS OF TRAFFIC, GEOGRAPHY AND ECONOMY

ABSTRACT

Modeling Internet growth is important both for understanding the current network and to predict and improve its future. To date, models have typically attempted to explain a subset of the following characteristics: network structure, traffic flow, geography, and economy. I will discuss a discrete, agent-based model, that integrates all these aspects. The model will be compared to empirical data. Network topology, dynamics, and (more speculatively) spatial distributions that are similar to the Internet.

*KTH EE/LCN, Stockholm, Sweden, pholme@kth.se

Keywords – Network tomography, passive measurements, TCP, performance, troubleshooting

Ernst BIERSACK*

TCP ROOT CAUSE ANALYSIS REVISITED

ABSTRACT

Throughput is the key performance metric for long TCP connections. The achieved throughput results from the aggregate effects of the network path, the parameters of the TCP end points, and the application on top of TCP. Finding out which of these factors is limiting the throughput of a TCP connection, referred to as TCP root cause analysis is important for end users that want to understand the origins of their performance limitations they experience, ISPs that need to troubleshoot their network, and application designers that need to know how to interpret the performance of the application.

The seminal work in the area of TCP root cause analysis by Zhang et al. (**Zhang, Y., Breslau, L., Paxson, V., Shenker, S.:** *On the characteristics and origins of internet flow rates.* In: Proceedings of ACM SIGCOMM 2002) has shown that for more than 50 % of the TCP connections analyzed, it is not the network bandwidth that limits the throughput but rather the application or mechanisms such as TCP slow start or too small a receiver window.

In this talk we will describe our system for TCP root cause analysis. The system takes as input bidirectional packet header traces captured at a measurement point anywhere along the path from source to destination and produces as output quantitative information about the limitation causes of TCPs throughput for each connection. We only analyze long lived TCP connections with at least 100 packets for which TCP slow start no longer dominates the throughput performance.

A great number of parameters determines the performance of a TCP connection, such as round-trip time, receiver advertised window, link capacities, available bandwidth, or version of TCP. We extract these parameters from the packet header trace and use them to derive the most likely root cause (or causes).

We distinguish three main classes of root causes: (i) limitations due to the application, (ii) limitations due to the TCP end-hosts and (iii) limitations due to the network.

For a single long lived TCP connection, it is very likely to observe different limitation causes during different periods of time. For instance, some Internet applications such as BitTorrent or HTTP 1.1 operate by switching between active transfer periods and passive keep-alive periods.

A major challenge is to detect those different periods and analyze them separately. For each connection, first the periods where the throughput is determined by the application are isolated. The

*Eurécom, Sophia Antipolis, France, erbi@eurecom.fr

remaining parts of the connection consist of one or more so called bulk transfer periods that are then further analyzed for limitations due to the TCP end-hosts and limitations due to the network.

Analyzing Internet traffic at packet level involves generally large amounts of raw data, derived data, and results from various analysis sub-tasks. Typically, the analysis often proceeds in an iterative manner and is done using ad-hoc methods and many specialized scripts. Instead, we have decided to use a Data Base Management System, PostgreSQL in our case, to manage the different data and implement the various analysis procedures.

We will not only describe our system and present some of the results obtained, but also discuss the difficulties encountered, relate our exerience using PostgreSQL, and indicate possibilites for carrying the work further.

More details can be found in **M. Siekkinen, G. Urvoy-Keller, E. W. Biersack, and D. Collange.** *A Root Cause Analysis Toolkit for TCP.* Computer Networks, 52(9):1846–1858, 2008.

Under `http://avband2.eurecom.fr/` we make available a service for performing TCP root cause analysis that is particularily interesting for users connected to the Internet via DSL, Cable, or WLAN hotspots.

Sonja BUCHEGGER*

LOCATION AND FAIRNESS IN SELF-ORGANIZED NETWORKS

ABSTRACT

Network nodes may experience large disparities in utility according to their location in the network topology, mainly attributable to differences in traffic load due to shortest-path routing. These disparities become more problematic in resource-constrained self-organized networks, such as mobile ad-hoc, peer-to-peer, wireless mesh or sensor networks, than they have been in traditional infrastructure-based networks.

The impact of node location has so far received relatively little attention, e.g. it is common practice to assume the random-waypoint mobility model in mobile ad-hoc networks, implying that over time node location will be evenly distributed. We are interested in the effect of location on node utility when this assumption is removed.

Applying insights from graph theory and social network analysis, we use centrality metrics and quantify the effect of location and several network topologies. We then use metrics and methods for equity taken from economics. Combining such metrics, we can understand better how choices of mechanisms and topologies impact the total network performance as well as its distribution over the network nodes.

As a concrete example of the general problem, we investigate how incentives for cooperation (such as payment or reputation systems for traffic relay in mobile ad-hoc networks) exacerbate or alleviate node utility disparities due to location. We show that location matters and that without location awareness, such incentive schemes can be unfair.

The generalization of the problem definition (of location-dependent fairness) and the methodology (of quantifying the distributions of both location and utility as well as evaluating the suitability of different notions of fairness) can be applied to several domains, such as evaluation of mobility models, strategic node behavior (location changes) in mobile networks, placement of access points in wireless mesh networks, of road-side units in vehicular networks, sinks in sensor networks, topology control of overlay networks, location-aware incentives for cooperation, and in general evaluation of fairness of networking protocols.

*Deutsche Telekom Laboratories, sonja.buchegger@telekom.de

Fabio RICCIATO*

A SNAPSHOT OF 3G MOBILE TRAFFIC

ABSTRACT

Third-generation (3G) mobile networks are becoming increasingly popular as access infrastructures to the global Internet. 3G networks are still undergoing a fast evolution phase, due to a continuous process of architectural upgrades, increase of network capacity, terminal evolution and tariff change. In such an evolving framework, traffic monitoring is the key to obtain and maintain understanding about the behaviour of the network and its user population.

In this talk I will report on our experience with monitoring an operational 3G network. The core of our research is focused on developing a concept for anomaly detection in support of the network operation process. We take a distributional change-detection approach. A number of traffic features are measured per-mobile terminal (number of connection openings downloaded packets, etc.), at different time aggregation time-scales from 1 minute up to 1 hour, from which time-series of feature distributions are derived. Building upon the notion of KL divergence, we have developed a change-detection scheme aimed at reporting "anomalous deviations" from what has been observed "in the past". This requires a preliminary understanding of the "typical" (normal) temporal characteristics of the distributions, to be achieved by an extensive exploration of the data at hand. The talk provides an overview of such exploration, enlightening the key findings about the characteristics and temporal (ir)regularities of feature distributions. The work is based on two large datasets, collected one year apart, covering several weeks of data from an operational network. Furthermore, we discuss the design principles of our anomaly detection system, and report on our operational experience.

*FTW and Univ. Salento, ricciato@ftw.at

Gábor VATTAY*

RECONSTRUCTION OF TRAFFIC AND TOPOLOGY FROM ACTIVE MEASUREMENTS

ABSTRACT

In this talk I concentrate on one question: Can active measurement based monitoring and management be a viable competitor of passive methods in a Future Internet? The advantage of active methods is that they raise limited or no privacy / security concerns. Data can be shared without anonymization. They can also be implemented without the collaboration of internet service providers. Yet, these methods provide key information about their networks. The disadvantage of active methods is their limited detail and accuracy.

We investigate two interrelated problems in detail: The efficiency of the active topology discovery and active network tomography.

We present new analytic results for the link and node discovery probability and for the number of links and nodes discovered by traceroute-like shortest path based methods. Our most important finding is that the link discovery probability in real networks is several order of magnitude different than it is predicted by the mean-field approximation. As a consequence, the number of discovered nodes grows linearly. We also give an analytic relation between the true and the discovered degree distribution. We demonstrate this in simulated growing complex networks, real Internet data provided by traceroute measurements done in the PlanetLab infrastructure. The result can also be applied to Peer-to-Peer overlay networks where the average number of Internet routers affected by the P2P traffic can be estimated.

Network tomography enables us to investigate the delay fluctuations in the discovered routers and links. We demonstrate that by now these measurements reached the accuracy of passive methods. Detailed analysis of the entire delay distribution is possible. It includes fitting of queuing models with long range dependent cross traffic and the determination of the Hurst exponent inside of the network. We discuss briefly the prospects of a new branch of statistics based on ensembles with network structure.

*Department of Physics of Complex Systems Eötvös University Budapest *and* Computer and Automation Research Institute of the Hungarian Academy of Sciences, vattay@elte.hu

AUTHOR INDEX

Biersack, Ernst 125
Bigos, Wojciech 91
Brauckhoff, Daniela 75
Buchegger, Sonja 127
Burkhart, Martin 75
Carle, Georg 57
Charzinski, Joachim 107
En-Najjary, Taoufik 9
Fu, Bingjie 25
Gauthier Dickey, Chris 41
Goerg, Carmelita 91
Holme, Petter 123
Klug, Andreas 91
Li, Xi 91
May, Martin 75
Münz, Gerhard 57
Papp, Gabor 41
Plissoneau, Louis 9
Ricciato, Fabio 129
Rodriguez, Pablo 5
Thiran, Patrick 3
Timm-Giel, Andreas 91
Uhlig, Steve 25
Urvoy-Keller, Guillaume 9
Vattay, Gábor 131